BACKYARD
LIVESTOCK

BACKYARD LIVESTOCK

Raising Good, Natural Food for your Family

Revised, Expanded Edition

Steven Thomas

Revised by

George P. Looby, DVM

Illustrations by Mark Howell
and Patricia Witten

The Countryman Press
Woodstock, Vermont

Published by The Countryman Press, PO Box 748, Woodstock, Vermont 05091.
Distributed by W.W. Norton & Company, Inc., 500 Fifth Avenue, New York, New York 10110.

Library of Congress Cataloging in Publication Data

Thomas, Steven.
    Backyard Livestock: raising good natural food for your family / Steven Thomas: revised by George P. Looby.
        p.   cm.
    ISBN 0-88150-182-4
    1. Livestock—Handbooks, manuals, etc. I. Looby, George P.
II. Title
SF65.2.T483 1990
636-dc20   90-15081
                                                                    CIP

Cover and text design by Virginia L. Scott

20 19 18 17 16 15 14 13 12

Discontent is the want of self-reliance.
Ralph Waldo Emerson

*For Meme, who fed the animals while
I wrote about them and without whom
none of this would be fun.*
and
*For Peter and Jane Jennison, for all
their help and encouragement over the
years and for their friendship.*
Steven Thomas

Contents

Publisher's Note

The original *Backyard Livestock* was written in 1976 by Steven Thomas, at a time when self-sufficiency was the goal of thousands of young people who moved to the country aiming to fill their every material need from their acres of farmland. Steven's book guided many of them as they raised their own food.

This 1990 edition of the book has been painstakingly revised and updated by Dr. George P. Looby, a veterinarian, whose advice reflects the careful thinking of the seasoned scientist, explaining how even today raising quality meat is a very doable project.

Dr. Looby has updated the language of the seventies and kept the good humor of Steven Thomas while also contributing immeasurably, particularly to the sections on Health and Breeding—his own additional specializations. We're proud to have Dr. Looby's expansion and enhancements of every chapter.

May your adventures with *Backyard Livestock* be successful!

Preface to the Revised Edition

This book was written to serve a a guide to those who have the desire (and the homestead) to start, in a modest way, a livestock project.

Livestock husbandry satisfies a basic human need, the need to have animals to observe and care for. There is something fundamental about having an animal as part of a family's basic unit. For many this is no longer possible. As our society has become increasingly urbanized, the ties that many people had to farms and farm animals have been lost.

Only three generations ago nearly everyone had those ties. That was a time when our farm population was a large part of our total population. Now that farm population is only a tiny faction of the total, the likelihood that individuals will have those ties continues to decrease. But this doesn't diminish that desire of so many to have a small plot of ground and to raise a few animals for their own use and enjoyment.

Another change that we have seen is our more sophisticated attitude toward personal health and physical conditioning. These have changed many of our ideas regarding the foods we eat. No longer is the consumer dedicated to prime cuts of meats, dairy products high in butterfat, and cholesterol-laden eggs consumed in great quantities. This change reflects the opinions of nutritionists and physicians regarding the types of food that we should eat. It does not mean that animal products no longer have a place on our dinner tables. Instead, it means the products presented should be modified to meet revised dietary recommendations.

As an example, for years the beef industry promoted the idea that the best cuts of meat were those with heavy marbling—intramuscular fat—and an adequate fat cover for maximum flavor and tenderness. Today the leaner cuts are favored.

These lean cuts can be produced far more economically and efficiently than the fatter prime carcasses promoted for so many years. The type of meat cuts now being advocated fit well into the backyard type operation explained in this book because the primary feeds needed are hay and grass and far less grain than the amount fed to produce prime cuts preferred before.

People thinking of developing an animal project should start slowly. There is nothing to be gained by starting with a large operation unless you are experienced.

You should remember that once you have decided to take the plunge and purchase animals, you have made a commitment. That commitment is to those animals. You must provide the time needed to care for them adequately. No matter how efficiently you operate, those animals will demand a portion of your day. For most species, that demand is made every day. You must make a commitment, too, to spend money on your animals to provide the care they require. Unless you are prepared to make this major commitment of time and money, you should devote your energy to a different project.

You must think, too, of your neighbors. We find ourselves in an increasingly restrictive environment. Zoning laws in most communities dictate what livestock we can keep on our premises. Be certain that you check local regulations to learn about restrictions and limitations before you make even a modest investment in livestock, housing, or fencing.

Today there are many people—but not as many as there once were—who yearn to purchase that "little place in the country" where they can keep the kind of livestock they fancy. Unfortunately, the "back to nature" attitude is no longer as prevalent as it was.

A new group has taken its place: These are the consumers who have become disenchanted with "factory grown" livestock produce. They have a strong preference for products raised under more natural conditions where the animals are not fed growth or antibiotic additives and are allowed to range freely or at least not grown under conditions of extreme confinement.

This segment of the consumer population offers a profitable market for anyone contemplating a modest livestock operation, since surplus products can be sold to them at prices that usually are higher than those paid for the products of more commercial operations.

It is enjoyable to have a hobby develop to the point where it can at least partially pay its own way. Given the right situation, it is entirely possible for a backyard livestock enterprise to develop its own cash flow.

In the more than a decade since the original edition of this book was published, there have been many changes, ranging from the care and treatment of animals to the tastes and eating habits of Americans. As the reviser of this book, it has been my task to update those parts that needed it and to add newer information that might not have been available at the time of the original writing.

Recognizing an interest in beef projects, we have added a chapter titled Beef Cattle, which emphasizes the use of a pasture to produce the lean grass-fed beef that is popular today. Try it if you have access to a pasture, enjoy raising animals, and want to eat the finest of beef. You may even save a few dollars.

But some things haven't changed. One of those is the love of animals and the urge to raise them, not so that we will eat better or less expensively, but to satisfy a longing that is deep within us. To those people I offer this book, like its original version, as a guide to success and deep satisfaction.

George P. Looby, D.V.M.
1990

Preface to the First Edition

The land is a mother that never dies.
Maori

The reports are ominous. Drought continues in the sub-Sahara; the Soviet Union has suffered yet another dismal growing year. Even now farmers in the American Midwest are feverishly harrowing their fields to prevent precious topsoil from being blown away by strong winds. There is already talk of another dustbowl like the ones that occurred most recently in the thirties and fifties. For those who are confident that technology will again bail us out, little comfort is given by the fact that climatologists see more such disastrous farming weather in our future than we've had for years—even centuries. They describe the exceptional growing weather we have enjoyed since the thirties as just that—exceptional. Such consistently ideal weather for growing has not occurred in the last thousand or more years, and it is not expected to return for a like number. It is not very comforting to realize that we have based our lofty estimates and stocks for the future (as well as seeds, soil science, and planning) on such atypical weather. How will we grow enough food for the future?

If the world's population were declining—even leveling off—there would still be cause for great concern. However, the population is increasing at a rate of 2% per year. As I write this (March, 1976) there are close to 4 billion people in the world. Every *month* that passes since I wrote this, the population will increase 7 *million*— almost a quarter of a million people per day! At the present rate of growth the population will double every thirty-five years. The population will increase 4 billion to 8 billion by 2011, and in the next thirty-five years it will increase 8 billion to a total of 16 billion in 2046!

Population pressures have resulted in the farming of marginal land and the overgrazing of livestock with the result that in drought years these lands may become desert. A United Nations study indicates that as much as 6.7 percent of the earth's surface may be man-made desert.

Increasing shortages and constantly escalating meat prices are inevitable because of drought, population pressures and higher feed prices. My answer to this problem, obviously, is to grow your own meat. To those of you who doubt your ability to do this, don't forget that it has only been in the last 50 years or so that much of the human race has given up growing its own food. Is it hard work? Is it time-consuming or just plain drudgery? Hardly. On the scale we're talking about—growing meat for one's own family—the investment in time and effort is miniscule compared to the rewards. To raise a pig, a couple of lambs, and some laying hens the daily time requirement would be less than half an hour. Your animals needn't cut down on your mobility. If you want to go away, have a neighbor take care of your stock; you can reciprocate when he goes away. We have baby sitters—why not animal sitters? And you cannot put a price tag on the pleasure you derive by doing it yourself, not to mention the superiority of your own products over their store-bought counterparts. It was enlightening for me, having grown up in New Jersey, erstwhile Garden State, to discover that a lamb chop is a naturally occurring item found along the backbone of a lamb, not a commodity wrapped in plastic, nestled next to Extra-Lean Ground Chuck in Aisle Six.

Frances Moore Lappé in her book *Diet For A Small Planet* calls meat animals a protein factory in reverse because a commercially raised steer, for example, consumes over 21 pounds of *humanly edible* plant proteins to produce just *one* pound of meat for human consumption. Therefore, critics say that the real answer to the world's food problems is to give up meat. Fine. Millions of people are vegetarians by custom or preference, but most Americans would sooner give up Apple Pie (and even Mother) before their meat. The answer is to grow meat more efficiently, using wasted food, pasture, or otherwise unused foodstuffs—and that, in part, is the aim of this book. Even Ms. Lappé admits that meat animals *need not* be wasters of food, that they can indeed be "protein factories." Ruminants (sheep, goats, cattle), because of bacteriological action in their stomachs, can convert low-quality feed into high-protein meat. They can graze secondary or unused pastures, vacant fields that lie fallow in wait for industrial parks, and even suburban lawns. The other animals and poultry we will discuss can make similar use of unused or wasted food or feedstuffs you can economically grow yourself to produce meat and other products for your family.

Let's face it: The United States is a wasteful nation. We throw away bottles, we throw away cans; using some pre-infantile logic we extol the virtues of a lighter that, still in working order, we can throw away without having to refill (Hooray! Another breakthrough!). Most of all, we waste FOOD—food that is humanly edible or could at least be fed to animals which in turn would produce meat. Perhaps you've never really thought about it, but consider how much food your family wastes: dumping a half-finished bowl of soup, throwing away that piece of meat that won't quite make a meal, throwing away glasses of milk, pieces of cake, etc. The best-fed animals in the world are America's garbage disposals. Think of the *tons* of food thrown away daily by schoolchildren, by supermarkets, by restaurants . . . your family. Just think.

The *Vermont Natural Resources Forum* reports that Dr. William L. Rothje, a

Harvard archaeologist, has been studying the behavior of Americans by examining their garbage. In Tuscon, Arizona, he found that the average person throws away 42 pounds of *edible* food per year—that's one-tenth of all food brought home! If 15 people saved their wasted food, they could raise a pig and grow almost 200 pounds of pork FREE. Take restaurants: it is their business not to waste, but they still throw away tons of food that is edible by animals (and in some cases humans). We save food scraps from one small restaurant, in one very small town in one of the smallest states—Vermont—and feed it to our pigs. Last year on those scraps alone we raised *three* 250 pound pigs that furnished us with over 500 pounds of delicious hams, bacon, sausage, and pork chops. FREE. On what would have normally been thrown away to rot.

Many of us have grown up in a time of plenty, free from hunger, and think nothing of waste. Others began their lives in a time of need and reason, "I worried about it then, but now I can finally afford not to." In both cases the result is WASTE. It is time for that to stop, and that is another aim of this book.

Right now, the "Backyard" concept is limited to those rural, semi-rural, or "exurban" areas where animals are allowed, or at least not dis-allowed. If enough people want it, restrictive suburban zoning can be liberalized. Animals are not carriers of disease; they are not filthy—only their owners are. The much maligned pig is one of the cleanest of all livestock and, if given the chance, might have taken better care of the Hudson River than we did. My sister and brother-in-law, living in suburban New Jersey, have about 50 pigeons, a flock of laying hens, and a horse. The neighbors complained at first, probably only because it was illegal, although they were never bothered. When they were offered the horse periodically as a free lawnmower, they were delighted. The fresh eggs were likewise welcomed. The fact is, if properly cared for and in sane numbers, animals can only be an asset. "But what will we do with their manure?" If you lived in the country the question would be "Where can we get more"—for the garden or to sell for extra cash. In time when gardens will be more of a necessity than a hobby, manure will be coveted. I think it's a new wave; it will be happening more and more. If you wake up some morning to the sound of a lamb bleating across the road or a chicken clucking as it scratches for food outside your window, be happy your neighbor cares. That's one more mouth off the food chain—better for you, better for everyone. In former lean times politicians promised "a chicken in every pot"; those campaigning in the near future will do well to promise a lamb on every lawn, chickens in every lot.

Many people have contributed their time and knowledge to help in the production of this book. The author would like to express his gratitude especially to the following: my editors, Jane Jennison and Marguerite Sheffield, for laboring over at times illegible typescript to make this work into a cohesive book; Henry and Cornelia Swayze, for their own information on sheep; Patricia Bryson, for her information on pigs; Bill Sumner, Windsor County (Vermont) Agricultural Extension AGent, for his advice on many aspects of this book; Drs. Jim and Steve Roberts of the Woodstock Veterinary Clinic, for many hard-to-find facts and their ongoing help with our animals; and all the people who took the time to send me feeding routines that they have found to be helpful in raising their livestock.

Steven Thomas
1976

Introduction

*Through want of enterprise and faith men are where they are,
buying and selling, and spending their lives like serfs.*
                                                Henry David Thoreau

I think the best advice to give people beginning to grow their own meat is to *start slowly*. Raising animals is a most enjoyable and satisfying experience, but it can be a nightmare if you get in over your head. We started a bit too fast: we had a couple of horses and some chickens; then we got a couple of pigs and some sheep, and before we knew it, the Vermont winter was almost upon us and we had no suitable shelter for our now pregnant animals. We did, somehow, manage to get our animals settled for the winter (our pig two days before she farrowed), but we vowed not to get any more animals until we had pens constructed and we fully realized how much additional time each animal would take. Table I.1 will give you a rough idea of how long each animal (in the case of poultry, a dozen birds; sheep, six to twelve head) will take of your time each day, summer and winter:

Table I.1: Approximate Time per Day to Care for Livestock

	Time/Day (Minutes)	
Animal	Summer	Winter
Poultry	5	7
Sheep	5*	10
Milk goat	15–30	15–30
Pig	5	7
Rabbit	5	7
Veal calf	7	10
Beef calf	5	10

*If you raise grass lambs that are contained by fencing you will have practically no time investment (except to check their general well-being) other than to provide water. If you grain them in the summer, figure an additional few minutes.

The difference in time required between summer and winter reflects the need to carry and thaw water and to clean pens (since animals will spend more time indoors in the winter months). The time you'll need to care for your stock will depend on how far you have to carry water, how well organized your pens are, etc. Don't let the amount of time scare you if you're figuring on keeping a number of different animals. If you have 3 pigs, a flock of chickens, and a dozen sheep, it probably won't take a full 7 minutes per pig, plus 10 for the sheep and 7 for the chickens in the winter for a total of 38 minutes per day. There is a lot of overlap. A big pail can be filled up to supply water for all the stock; pigs are fed the same ration, near each other or in the same pens. Above all, with experience you will organize your barnyard, or yard, so your time will be used most efficiently.

The animals I have included in the book were chosen for ease of care, for relatively small space requirements, desirability of products and, most importantly, their ability to forage for their own food or to make use of low-protein food, otherwise wasted foodstuffs, or the products of other animals.

The most important feature of this book, in my mind, is the section *Supplementing Commercial Feed* found under the *Feed* section. Here I have attempted to give you some independence from the feed companies by pointing out other feeding routines you can implement yourself. Let's be pessimistic and assume there'll be grave problems feeding the world in coming years . . . or, if you will, let's just be practical: feeding your animals without buying commercial feed is a cheap way to eat! This is just a beginning. I have collected some methods, either from my own or other people's experience, of alternative feeding. I should warn you, though, that I have not personally tried *all* the supplemental feeding routines, so use some caution when implementing them. Best of all, use them as a starting point in developing your own routines. I make this appeal: If you know of, or develop any effective and economical feeding routines for any of the animals in this book, please send a complete description to:

> "Feeds"
> c/o The Countryman Press
> Box 175
> Woodstock, VT 05091

and you will, at least, receive acknowledgement in any future editions of this book. I believe this is an exciting prospect and hope we can gather and share such information.

I have tried to organize this book, with its chapter divisions and subdivisions, to make it easy to refer to any aspect of livestock care. Most chapters are divided into nine major sections, listed below with a description of each:

1. Breeds: I have not tried to acquaint you with all major breeds within a species; however, the most common and useful for a backyard operation are discussed, descriptions are given, and in some cases sketches are provided.

2. Purchase: The how and when are discussed as well as the whys of purchasing a certain animal. Features to look for, and avoid, are offered for each animal. This also includes information on how to care for the young animal. In the case of poultry, hatching as well as brooding facilities and methods can be found. To get a first-hand idea of what a good specimen of a given species looks like, go to 4-H clubs, livestock shows, or county fairs and inspect prize-winners.

3. Housing: Simple shelters for summer raising as well as more elaborate winter quarters are described. Also included are details of fencing and other methods of confining your stock.

4. Equipment: Feed dishes, water systems, and other day-to-day needs are discussed and simple plans shown to enable you to build your own. More specialized equipment for breeding and health care are covered in their appropriate sections.

5. Feed: This section is divided into (a) Conventional Feed and (b) Supplementing Commercial Feed. The first subdivision covers those feeding practices using commercially purchased feed, hay, or the like. When buying your feed, be aware of the going prices. Feed stores in one area often charge sizable differences in prices for the identical or comparable feed. Feed is usually cheaper in summer. Whenever the prices takes a sharp drop, buy in quantity if you can afford it and can store it properly. The proper watering of stock cannot be stressed enough. Water should be available, thawed, *at all times.*

Supplementing Commercial Feed has been mentioned above. In addition, topics in this section include mixing your own feeds, free or very cheap sources of feed that can be substituted for commercial feed, and hints on how to stretch your feed dollar and make your meat cheaper. If you focus on collecting supplemental feed—and make a real commitment—you will be amply compensated by cheaper meat costs.

6. Management: This section has three subdivisions: (a) Routines, (b) Handling, and (c) Predators. It also contains information that does not fit in any other section.

"Routines" deals with the management of your animals. Suggestions are made as to when to raise particular animals, in what numbers, and how to establish the most efficient and money-saving regimen of raising an animal for your family.

"Handling" involves just that—the correct (and easiest) ways to catch, transport, or carry particular animals.

"Predators" points out the most common animals that will prey on your stock and how to prevent or eradicate them. As a rule, a good barn cat will do a good job protecting your investment in animals and feed.

Keeping accurate records is an important facet of any livestock raising operation. Those "bargain" eggs you think you're getting may not turn out to be such a big deal when you calculate how much feed is going into the production of each dozen eggs. It is easy to fool yourself if you don't keep records. You might remember only four bags of feed you bought for your pig when you actually used eight—keep track. I like to keep a separate book on each animal (or flock, herd, whatever) to see how they pay off in the long run. Depending upon your aims—

profit, break-even, or just good eating—you can make an accurate evaluation of how each animal is producing and decide on the most efficient routines for your family. These records should also contain other pertinent information, such as breeding dates and outcomes, butchering weights, overall cost per pound, and the like. These records are also fun to look back on over the years to see how efficiently you've learned to run your operation. It's also nostalgic to look back at how cheap feed used to be. (They are also necessary if you are filing income taxes as a farmer.)

You will discover in time that the concept of backyard livestock makes for better stock. In raising animals, smaller numbers are best. Diseases are reserved, for the most part, for animals that are overcrowded and lack proper sanitation, housing, and ventilation. You shouldn't have any such problem with a backyard flock. In larger operations an owner simply doesn't have the time to check each animal daily and separate those animals that aren't competing well for food. In a small operation you will have few, if any, low grade animals because you can offer individual attention. If one of three pigs is being bullied you can, without much bother, make another pen for it or feed it at some distance from the other two. A large-scale owner must accept such poor competitors or take a financial loss and cull him. I can see it with our sheep. When we had a half dozen, they all got their share. As our flock grew, and separate feedings and other "coddling" became impossible, "poor competitors" became apparent. I had to take measures to ensure they all got their share or resign myself to a a few unthrifty sheep. In short, our animals are superior because I can watch each one carefully and spot any "bullying," signs of disease, or other problems early and correct them before there is any permanent damage. In a larger operation, this is far more difficult and time consuming.

In managing your animals, try to spend some time with them every day. The first sign of disease is often a drop in feed consumption; and if you're attentive, you'll spot this right off. By standing and watching our animals as they run in the barnyard or graze, I can spot even the slightest abnormalities: foot problems, lameness, cuts, or other unusual behavior. Spend some time with them and, above all, *be nice to them.* They are furnishing you with a most precious product. Pet them, talk to them, scratch them—never be mean to them—and they will pay you back.

7. Breeding: This section has details on breeding practices. It includes information on when and how to breed, gestation, feeding during pregnancy, equipment for breeding and birth as well as weaning and castrating.

8. Health: The key to guaranteeing the health of your animals is *prevention.* It is repeated so often throughout the book that you may get sick of it, but it does bear repeating. If you furnish your stock with proper amounts of food and water, clean quarters, and fresh air and sunshine you will, unless there's a local epidemic of an infectious disease, have a disease-free operation. This section lists the most common afflictions such as parasitic infestations (worms), their prevention and treatment. A disease table for each animal including major afflictions, their causes, symptoms, prevention and treatment, is in Appendix B for easy reference. These charts should aid your quick diagnosis and treatment of major diseases. If you are really serious about raising livestock, you would do well to purchase one of the veterinary manuals listed in Appendix F.

9. Butchering: Complete butchering information is given for all poultry and rabbits. The procedures for the remaining animals are too detailed to be given adequate coverage in this volume. If you watch them being butchered once or twice and use a book on butchering, you should be able to master the procedure yourself.

You might be able to sell or barter some of your surplus meat to further reduce your costs, but check state laws first. Many states require meat for sale to be butchered only in state-registered slaughterhouses. You do, however, have the additional option of selling the animal alive and having the purchaser arrange his own butchering.

The chapter, "Grow Your Own . . ." is designed to be used in conjunction with the "Supplementing Commercial Feed" section and gives you an introduction to growing your own grain crops, pasture, and making hay and silages. There are also references in Appendix F you can use to obtain additional, detailed information on growing your own feedstuffs.

I have attempted, within sane limits, to make this a one-book reference. I hope it will suit both the person who wants to raise a few laying hens and a lamb and a pig to those who wish to get more deeply involved, raising animals on a fairly large scale, arranging their own breeding, and doing their own castrating, butchering, and the like.

We are lucky to have livestock and poultry that consistently grow fast, live long, are disease-free, have large and healthy litters, etc. If anyone were to ask me if there is any sure way of raising superior livestock, I'd have to say yes. Simply: *do it by the book* (I'm not being immodest—you needn't use this book—do it by *any* book). Do what you're supposed to do. Furnish your stock with plenty of fresh water, a proper and clean pen, and the correct amount of feed, and that's it. The stock I see having problems usually aren't getting water or their owners are taking other shortcuts. Such people suppose that, for example, by cheating on feed a bit, the pig will grow anyway and be cheaper. Not so. If it says to feed your nursing pig ten pounds of feed a day—*do it*. Sure, she'll probably make it on seven pounds; maybe won't even lose any pigs, but she'll be run down and have a shorter productive life. Then see if you get a reputation for having good piglets!

Finally, I don't profess to be an expert. Not by a long shot. Since I have finished this book, I have learned countless things I should have included. Many people know more about animals than I do and have raised them for more years, but I guess they didn't want to write about it. There are no "experts," you'll learn every day no matter how long you've raised animals. In that sense it's presumptuous to think that this book is complete. Refer to other books in the Catalog of Resources, pump other people for information, keep your eyes and ears open, and you'll be surprised at how fast you'll learn. It may have taken you eight years to learn your multiplication tables, but you'll learn about farrowing the day your sow is due.

1

 POULTRY

Chickens

Of all the animals that anyone could choose starting out in a backyard setting, the most suitable one is the chicken. The start-up costs are low and the chicken's needs are modest. Your own time commitment should be low, too, unless you just enjoy watching hens scratching in the yard.

Even though eating eggs is no longer recommended at the levels previously promoted, eggs are still enjoyed at lower levels of consumption. There is no restriction on eating egg whites.

Chicken is increasingly advocated as a major portion of the total meat intake in our diets. Raising chickens at home for this purpose is a most practical consideration.

BREEDS

Almost 200 different breeds of chickens are listed in the *American Standard of Perfection*. If you peruse a poultry catalog, you will be amazed at the many unusual and downright bizarre breeds of chickens, but for our use we'll classify chickens into three groups (1) "egg" birds; (2) "meat" birds; and (3) "dual-purpose" birds (Table 1.1, Figure 1.1). The egg birds are small and are excellent egg producers, but because of their size they are rarely raised for table use. Meat types are large, broad-breasted birds that convert most of their feed to meat and are usually poor egg producers. Dual-purpose birds incorporate the qualities of both egg and meat birds into one and enable the homegrower to make an enjoyable meal out of any egg producer that is foolish enough to stop laying.

If you choose to use one of the egg-laying breeds, remember that the majority of the breeders have so refined their egg-laying characteristics that these chickens will lay well for only one or two seasons.

Of course, your choice of breed will depend on the needs of your family. Once you taste rabbit, you may simplify selection by getting a few layers and relying on your rabbits for meat, but that's another consideration altogether. In dual-purpose

Table 1.1: The Three Classes of Chickens and Common Breeds Within Each

Egg Birds	Meat Birds	Dual-Purpose
Leghorn	Cornish	Plymouth Rock
Minorca	Orpington	Rhode Island Red
Ancona	Australorp	New Hampshire
Blue Andalusian	Brahma	Wyandotte
Bantam	Cochin	
	Langshan	

birds I have found Plymouth Rocks superior to New Hampshires and Rhode Island Reds in meat quality and about equal in egg production. The White Plymouth Rocks dress out more attractively than the Barred Plymouths with no dark feather shanks left in the skin, but you can often sell the neck feathers and skin* of the Barred Rock roosters to sports stores for use in fly tying.

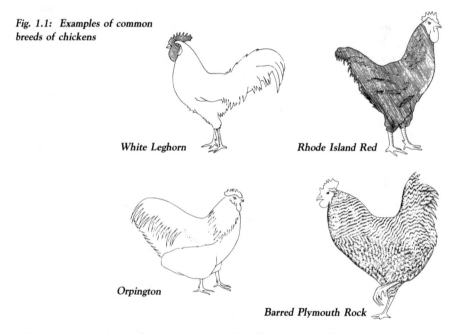

Fig. 1.1: Examples of common breeds of chickens

White Leghorn

Rhode Island Red

Orpington

Barred Plymouth Rock

By experimenting and trying other kinds of birds, you will find out which are best suited to your needs and taste. Most commercial egg and meat birds are hybrids. Our first birds were a Bantam—Rhode Island Red cross. They incorporate the docility and egg-laying ability of the Bantam with the somewhat larger eggs

---

*To prepare it, take the skin off the neck with feathers intact, and salt to preserve it.

and better meat quality of the Rhode Island Red. For the ambitious poultry raiser, cross-breeding opens up an interesting opportunity for experimentation. I remember a story in the news a few years ago about an Australian teenager who, by crossing at random different breeds of chickens, came up with a bird that matured to 26 pounds and laid one-pound eggs! Panic-stricken poultry companies offered a million dollars for the bird so they could *destroy* it and with it any competition for their rubbery chicken and mass-produced eggs. Since I have heard nothing else about it (and have yet to see 25-pound chickens in the supermarket), I can only assume the boy accepted their offer and now lives a life of leisure munching buckets full of Kentucky Fried Chicken.

PURCHASE

To decide on the number of chickens you'll need, take into account your egg and meat consumption and determine whether the birds are intended for home use or whether you plan to defray their feed costs by selling surplus eggs. As an aid in helping you decide the number of birds you need for your particular situation, figure a good layer in her first year of production will give you 200 to 220 eggs. The first year of production is the best year, and each succeeding year a hen produces fewer eggs. That decrease depends on health, breed, age, genetic makeup, and, most of all, the care you give her. Generally, you can depend on almost one egg per day per bird until she goes into molt (see *Management* section), when production usually ceases. As far as meat is concerned, again depending on breed, you can expect a bird to dress out at between 2 and 5 pounds at the age of 5 or 6 months. If you plan to market eggs in your neighborhood, you can buy as many as you have space for and can handle financially, but don't undertake more than you think you can market. However, if you are just starting out, it is best to fill your family needs and worry about selling eggs and/or meat at a later date.

Once you have decided on your breed and quantity you are ready to make your purchase. You will have the option of buying mature birds (i.e., birds of production age or older), day-old chicks, or fertile eggs for hatching. Your initial plunge would probably be easiest if you purchase ready-to-lay pullets from a poultry farm. This frees you from the problem of hatching or brooding chicks, plus the wait of 4 to 6 months for your first eggs, and the discouragement of losing them at this young age due to disease or beginner's ignorance.

Mature Birds

If you take this option you are ready to gather farm-fresh eggs the next day (or as in our case, on the ride home). Often, depending on your source, it may be cheaper to buy mature birds rather than feeding them for the 5 months it takes for chicks to begin producing. You might purchase them from a nearby poultry farm, a livestock auction, or from an overstocked neighbor.

The disadvantages of buying mature birds are that you will probably be limited in your choice of breed, and if you're not careful you might buy poor laying or nonlaying birds. Table 1.2 outlines characteristics to look for in choosing a good laying hen. This can be used both in purchasing your birds and in culling birds from your flock.

There is still a bit of folklore that suggests that you should have a rooster in your flock to maximize production. This is not the case. The only reasons for a rooster are to have a readily available source of fertile hatching eggs, or if the sight and sound of a rooster rounds out your idea of what your place ought to be. Recent news releases tell sordid stories of roosters introduced into suburban neighborhoods and their unhappy fate.

Table 1.2: Characteristics of Good and Poor Layers

Characteristic	Good Layer	Poor (or Non-)Layer
General health and appearance	Alert, well proportioned, active	Deformed, listless, weak
Eyes	Bright and alert	Dull, listless
Comb and wattles	Thick, smooth, and bright red	Pale, shrunken, dry, scaly
Vent	Large, moist, oval; white or pinkish	Small, dry, round; yellow
Pubic bones	Flexible, wide apart; at least three fingers in width	Hard, two fingers or less apart
Abdomen	Soft and pliable, thin-skinned	Firm, thick-skinned
Pigmentation	Pigment bleached from vent, earlobe, beak and shanks (legs)	Yellow pigment in these areas
Plumage	Bright, smooth, clean	Dull, matted, dirty

Day-Old Chicks

Buying just-hatched chicks enables you to have a wider choice of breeds, to watch them grow, and have that inexplicable joy of finding that first walnut-sized egg 5 months later. Chicks are usually sold in three classes: 95 percent pullets, as-hatched, or straight run (not sorted by sex), and cockerels. You will pay the most for pullets, and the least for cockerels, with as-hatched falling in-between.

Try looking for a hatchery in your telephone book, or you can surely locate one in the back of a farm-oriented publication. Another source of day-old chicks is a feed store. Often in the spring as promotion for its feed, a feed store will give away 25 cockerel chicks of a meat type with the purchase of a bag of feed.

Chicks should be ordered only from flocks that are certified to be pullorum-free. Pullorum-infected chicks (see *Health* section) are often unthrifty and mortality can be quite high. Chicks that are shipped are especially vulnerable, so deal with certified hatcheries. Unless the hatchery is local, the chicks will be sent through the mail. Don't shudder at this—for some reason the never-reliable mail service excels when delivering day-old chicks. (Perhaps the answer to poor mail delivery is to enclose a chick with every piece of correspondence.) After hatching, a chick has enough nourishment for two days, which is plenty of time for delivery. You will be notified by the hatchery a day or two before your order is shipped, and miraculously the next day your post office will be filled with the sounds of life. Your next step, *immediately*, is to put them in a brooder.

Brooding is the process from birth to about 6 weeks of age whereby the chicks are kept confined (but not crowded) in gradually decreasing heat, with plenty of food and water and, above all, strict sanitation. It is the artificial process similar to the natural brooding you see as a mother hen stands in the yard with a clutch of chicks huddled beneath her.

The basic requirements in any brooder setup are:

Adequate space. One-half square foot per chick up to a month of age and one square foot thereafter until taken from the brooder.

Correct heat. 90°–95° the first week with a lowering of five degrees each week until 55° or the outdoor temperature is reached, whichever comes first.

Adequate food and water.

Adequate ventilation without drafts

Must be predator-free

I have seen much more elaborate recommendations, including those suggesting infrared heat lamps or gas-fired heating elements; but as long as the basic requirements are met, there should be no problem. For small groups of chicks we have used a cardboard box with a light suspended in it and placed it in our mudroom. For larger numbers I built a frame in the corner of an unused room and placed an old screen door on top. (The upper half of the door was screened and the lower half wooden.) I placed a 60-watt light bulb in the lower end and allowed the chicks to regulate the heat themselves by moving closer to or farther away from the light. Their food and water were placed in the screened end. The screening on top let in light and also allowed for good ventilation without being drafty. Another brooder for a large number of chicks is the hover type. A hover with an infrared lamp can either be suspended from the ceiling or set up off the floor with blocks so that it allows the chicks free access to the heat. Heat requirements are the same as listed above and the chicks can be allowed to use it free choice. Each chick should be allowed 7 square inches of hover space.

Litter with large pieces, such as wood chips or peanut hulls, are preferred as they help maintain even floor temperature, absorb moisture, are far less dusty, and allow larger pieces of manure to fall through.

In warmer weather, it is essential to check the temperature in the brooder to make sure it doesn't get too hot. During a particularly hot and humid day, turn the light off until the temperature drops in the evening.

Too much light and heat, plus overcrowding in a brooder, can result in cannibalism because of the natural tendency of the chicks. They will start pecking at each other, usually around the eyes and the vent; and, when blood is drawn, all will turn on that particular chick and peck it to pieces; then turn to the next unlucky one.

To prevent this, have the chicks debeaked before you buy them. This involves cutting a quarter-inch off the top beak soon after hatching. This service is almost always available from hatcheries. If you can't buy debeaked chicks, cutting back on light, heat, and overcrowding should stop the problem. Some people recommend using a light bulb no brighter than 15 watts, but we've always used 40 watts or higher.

At 4 weeks you should set a roost an inch or so off the floor in the brooder. It can be a length of sapling or a piece of half-inch pipe. At 6 weeks, when the young chicks are completely feathered out and if the weather is settled, the chicks can be taken from the brooder.

It is not good practice to mix young chicks with older birds. "Pecking order" is a phrase not arrived at by chance. Putting young birds with a group of older birds starts an adjustment period where the older birds dominate the activities of the young ones, especially at the feed trough. During the period of rapid growth, the young birds should be able to eat all the feed they can without fear of harassment by the older hens. Continually being pecked at is a very real stress that should be avoided.

The housing needs of young chicks are modest and, in good weather, their housing need not be elaborate. A simple shelter fashioned out of two pieces of weather-resistant plywood for a roof and with chicken-wire sides will serve nicely. Young birds routinely used to be placed on grass range in this type of shelter and were allowed to run on range until they approached laying age. That system is too labor-intensive for today's mega-chicken operations but remains an option for small-scale ones.

Regardless of the size of your operation, good sanitation is essential. Your brooding area should be cleaned and disinfected after the chicks are removed. Similarly, the area where the layers are kept should be cleaned when their laying careers are completed and before a new group of layers is moved in.

Hatching Eggs

This can be the cheapest method of starting a flock if you build your own incubator (see Appendix) and get fertile eggs from a neighbor. Having a broody hen do the work for you will free you from incubation or brooding problems, but you can't make a hen broody at will the way you can plug in an incubator when you want to hatch a few eggs. Details on hatching eggs can be found in the section on *Breeding* later in this section.

HOUSING

If you follow to a T plans in poultry books or agricultural bulletins, you will end up with a lovely house that could conceivably be used for a guest house if you ever sell your birds. It will cost accordingly. Use such plans for reference only, and with a little ingenuity and resourcefulness you'll build a good coop for a fraction of the cost.

The most important requirements for a good coop are:

Adequate floor space (at least 3 square feet per bird)
Good ventilation without drafts
Adequate lighting
Safety from predators

It is quite possible to make use of what you have around your property. One of the beauties of chickens is that they demand very little. If you live on an old farm, chances are there is an old milk-cooling shed. In two different places we have lived we have made use of these old sheds. They convert to fine chicken coops in a few hours (at the most) and will comfortably hold up to two dozen birds. Other potential coops include old toolsheds, erstwhile outhouses, or a corner of a garage or barn. If you keep an eye out you might notice an old shed on someone else's property. Possibly the owner will part with it for free (or a modest fee), and if it's movable, you're in business.

It is desirable to face the coop into the sun—to the south or southeast. If you are putting your chickens in a barn or other less well-lit location and are not allowing them to run free outdoors, you must supply artificial light. Chickens need 14 hours of light per day to lay optimally. Because laying hens give off a large amount of moisture, you should provide good ventilation. This can be a door that is left open (except at night) or windows that can be opened. Screened vents along the eaves of the roofline are good if they can be closed off temporarily during very cold weather.

You should have 1 square foot of window space for every 10 feet of floor space. Again, if you are making use of a preexisting structure you will have to make do with what you have. The windows in one milk shed we converted were for the most part broken, so rather than purchase new panes, I nailed chicken wire over the window and we covered it with clear plastic for protection during the winter. The chicken wire also serves to keep out the predators. Depending upon your area, predators—dogs, hawks, foxes, raccoons, and fisher cats—can lay waste to your flock unless you take measures to keep them out of the coop. Most important of all, close your flock in the coop at night. Since most predators are nocturnal, we allow our chickens to range during the day, and when they go to their roosts in the evening, we simply close the door.

However, there are often good reasons for not allowing hens to roam free, and instead to have a small yard attached to the coop where the hens can be outside throughout most of the year. Hens that are allowed to run freely throughout the gardening season will arouse the ire of the resident gardener. When the garden is started in the spring, chickens are attracted to the freshly turned soil and newly planted seeds. During this time they are better off confined to quarters so they won't end up in hot water.

For flooring we use sawdust because it is cheap and easy to get. You can use wood shavings, peat moss, straw, peanut hulls, or ground corncobs. Start with 6 to 8 inches and, instead of cleaning the coop out periodically, let the litter build up, stirring it occasionally, taking out only the soggy spots and adding more litter every month or so. One method of handling litter is to feed scratch on top of the litter. This encourages the hens to stir and aerate the litter.

You can avoid an unnecessary buildup of droppings on the floor by using one of two common systems. Both will keep the litter in good condition for a longer period of time. One system is to place a platform under the roosts where the droppings will fall onto it. The accumulation can be scraped off and placed in the garden or on the compost pile. (Remember that chicken droppings are very high in nitrogen and may stimulate unwanted growth of some plants.) Another system is to construct a boxed-in pit arrangement under the roosts. This allows you to let the manure build up for longer periods of time.

We usually clean out our coop once a year in the springtime. At that time we have over two feet of litter and it goes right into the garden because it is one of the best fertilizers we've ever used. After cleaning out the litter, wash the coop with a disinfectant.

Your coop will need roosts, located away from windows or drafts, allowing at least ten inches of space per bird.

Fig. 1.2: Two different styles of roosts

Fig. 1.2a: Parallel Roosts

Fig. 1.2b: "Stairway" Roosts

You should set them at least two feet from the floor of the coop and space them at least a foot apart. They can be placed on the same level with each other (Fig. 1.2a) if you have the room (this allows for better heat retention by the birds in the winter) or diagonally up the wall like treads in a stairway (Fig.1.2b). Two-by-two lumber is fine but straight saplings about 1 to 2 inches in diameter are just as good and cost nothing.

To keep the flock under control but allow the chickens to be outdoors, build a run (Fig. 1.3). This allows the chickens limited range as well as opportunities for dust baths and fresh air and sunshine. The run should be attached to the coop so that the chickens are free to leave and enter it. The sides should be 6 to 10 feet high to prevent the birds from "flying the coop." You can get by with 4-foot sides if you're housing heavy meat birds that can't get too far off the ground or if you clip a number of the primary feathers on the wings of your chickens to short-circuit any dreams of escape. It can be framed with two-by-fours or saplings and covered with chicken wire. Either bury the bottom edge of the wire, or place logs around the perimeter of the run to keep predators out.

Fig. 1.3: A chicken run

4'-8'
SEE TEXT

EQUIPMENT

You will need nest boxes at the rate of 1 per 4 layers (Fig. 1.4). They should be placed in a draft-free area in the back of the coop or wherever they will be least liable to distraction from traffic in and out the door. They should be 14 inches square and a foot deep. They can be filled with wood shavings or straw, and this nest material should be changed frequently to prevent soiled eggs. If space is at a premium, they can be stacked as long as they are not out of reach of the hens. I have a cover on the boxes since the birds seem to like it, and also to discourage roosting on the edge of the boxes with the resultant manure in the nesting material. If, because of the top a hen has trouble entering the nest, secure a small dowel in front of the box to allow the hen to fly up to it and then make her way into the nest.

You will also need feed and water dishes. They can be purchased fairly inexpensively, but so far I have resisted that urge and my homemade/makeshift feeders have done quite nicely. For feeders, the prerequisites are that the chickens be

Fig. 1.4: A nest box suitable for 8 hens

unable to scratch in them, hence wasting and contaminating feed, and that there is at least 4 inches of space per bird at the feeder.

A simple trough feeder with a reel can be built. If scratching in the feed is a problem, covering the trough with chicken wire so only the beaks can poke through is the answer (Fig. 1.5).

Fig. 1.5: A trough-type feeder. The length of plastic pipe on the steel rod will spin if stepped on, preventing birds from roosting on it and contaminating feed.

For a waterer I use an old kitchen pan about 4 inches deep and 8 inches in diameter. I made a little platform to raise it off the floor so that litter is not kicked into it. During cold weather drinking water must not freeze or there will be a drastic reduction in egg production. I favor hanging a 60- to 70-watt light bulb over the waterer to prevent freezing. This works at the coldest Vermont temperatures, is much cheaper to buy than an immersion water heater, and draws considerably less electricity. In a small coop this light can also be used to supply auxiliary light to maintain the 14-hour days needed for maximum production.

FEED

The diet we have developed for the chicken probably comes closest to that elusive perfect diet that Mother Nature had in mind. With it, healthy chickens for eggs and meat can be raised quite economically on commercial feed alone. However, people wishing greater self-sufficiency can enjoy substantial savings depending upon the time they wish to spend growing food.

To avoid problems during the chicks' first six weeks of life, the most critical period, it is best to feed a commercial starter mash with a 20 percent protein content. At no other age is the furnishing of ample supplies of fresh food and water more important.

In my brooder setup, I feed and water the chicks away from the light. For the first few days of their lives, I spread some newspaper or heavy paper over the litter and sprinkle feed on that, so the chicks learn the difference between their mash and litter.

Thereafter, make clean food available as free choice in chick feeders that you buy or build. (Build the same way you would for larger birds, but proportionately smaller.) I have built troughs on the side of the brooder using pieces of lath, which is even simpler and less expensive. These were 2 inches wide, 2 inches deep, and as long as was needed to accommodate the chicks. The narrowness and closeness to the wall of the troughs made it difficult for the chicks to stand in them and scratch out the feed. If they do, cover the trough with chicken wire.

At first allow 1 inch per chick of feeder space, increase to 2 at 3 weeks, and allow 3 to 4 inches per chick at 6 weeks. The greatest saving you can make in feeding is never to fill their troughs more than half full with feed. If you overfill it, they will waste tremendous amounts of feed (and money!).

You can buy a chick waterer, but I have always got by with a small bowl. I have used this method for up to 25 chicks but they do soil the water quickly, so it is essential that you check the water frequently and refill and clean the bowl if necessary. Allow a half-inch of watering space per chick.

Older Birds

You should buy commercial growing mash for chickens 6 weeks to laying age and laying mash for chickens of laying age and older.

With older birds, as with chicks, have the mash available at all times, and don't overfill your feeders or large amounts will be lost in the litter. A pelleted form of mash is available, and chicks waste less of it. If mash is spilled from a feeder, chickens will retrieve very little of it; with pellets any spilled feed will be visible and cleaned up.

Ranging

I believe in letting chickens "range" as freely as possible during the day, or at the very least, of building a run if ranging is impossible due to predator problems, land restrictions, or the difficulty of finding eggs laid while ranging.

Why exactly is ranging so advantageous? First, the savings on feed is tremendous. I have read that the saving on feed averages 20 percent; but in my experience it has been much greater: 50 percent or more. I feed my flock in the morning as I free them from the coop. They may make a few perfunctory pecks at the mash but are so eager to get out the door and to the earthy delights that await them that they usually run right past me and my grain scoop. They eat grass, bugs, grit, and whatever turns a chicken on. Also, which is important for us and those of you with other stock, they clean up after the other animals. With free-ranging chickens I guarantee you will have virtually no feed waste. Our chickens clean up after our sloppy horses, grab a few mouthfuls from our obliging sow, and spend a lot of time with their favorite (and best-fed) animals, the feeder pigs. We feed our feeder pigs almost totally on high-quality scraps from a local restaurant. After I empty the feed bucket into their dishes, I clean out the residue in the buckets (maybe a cup or two) with a hose and dump it on the ground. The chickens, admittedly among the dumbest barnyard creatures, have learned to come running from all corners of the farm at the sound of the hose, and they devour every morsel that is left. In the summer our chickens hardly touch their mash and get along famously on animal "leftovers" and their outdoor natural diet.

Scratch Feed

Scratch feed is a nonmash foodstuff that is fed in whole or cracked form by throwing it in the litter or on the ground. The formulas vary widely with some common ones listed in Table 1.3:

Table 1.3: Feed Formulas

Maryland		Louisiana		Washington		Michigan	
Corn	40%	Cracked corn	40%	Wheat	50%	Yellow corn	50%
Wheat	40%	Wheat	30%	Oats	50%	Wheat	50%
Oats	20%	Rice	30%				

You can save by growing your own components for one of the above mixtures or you can get by with a scratch feed of whole corn alone or corn with some sunflower seeds added. The grains should be cracked for chicks 6–10 weeks old and may be fed whole thereafter. Feed lightly in the morning and more heavily at night. I recommend spreading it in their litter (remove or cover water bowl and mash feeder or they will be contaminated as the chickens scratch for feed). It will also help aerate the litter, keep it drier, and expose some of the valuable nutrients in it.

I don't use scratch feed as a regular component of my feeding program. But as Table 1.4 shows, it can compromise up to 50 percent of the feed program, and if you can easily grow it yourself, it is worthwhile considering.

Table 1.4: Suggested Feeding of Scratch Feed

Age (weeks)	4	6	8	10	12	14	16
Mash (percent)	100	95	90	80	70	60	50
Scratch (percent)	0	5	10	20	30	40	50

Supplementing Commercial Feed

You can supplement the diet of chickens confined to coops or a run with freshly cut greens (alfalfa, clover, *fresh* lawn clippings, cabbage, kale, Swiss chard, and beet tops). During the winter, fine legume hay, clover, and alfalfa are good sources of feed as are sprouted grains.

A simple mix-it-yourself mash, many of whose ingredients you can grow or procure yourself, is shown in Table 1.5. A corn sheller and grain mill which can be ordered from farm catalogs is invaluable. If you do it in large enough quantities, it may be economically feasible to have a local grain mill do your grinding.

Table 1.5: A Simple Mash Mix

	Pounds per	
Ingredient	100	Ton
Yellow corn meal	60.00	1200
Wheat middlings	15.00	300
Soybean meal (dehulled)	8.00	160
Fish meal (65% protein)	3.75	75
Mean and bone meal (47%)	1.00	20
Dried skim milk	3.00	60
Alfalfa leaf meal (20%)	2.50	50
Iodized salt	0.40	8
Ground limestone (38% calcium)	6.35	127
Totals	100.00	2000

Source: *Raising Poultry the Modern Way* by Leonard Mercia. Charlotte, VT: Garden Way Publishing Co., 1975.

If so inclined, almost anyone can raise corn and possibly wheat, soybeans, and alfalfa. These constitute the bulk of the feed. Fish meal should be easy enough to make yourself (for that amount you need only be a marginal fisherman); meat and bone meal can come from the offal and bones of other livestock you butcher. If you have a dairy animal, your surplus milk can go into the feed. Even if you have to buy some of the ingredients, you can still save. One note: it *is* important to grind the feed into a mash, or the chickens will tend to pick out what they like, leave the rest, and not get all the benefits from the mixture.

For the number of chickens I have, I wouldn't think of going to the trouble of mixing my own feed. But for others it may be worthwhile. I depend more on ranging, food from the litter, and having our flock clean up after our other animals. Not surprisingly, this takes less of my time than raising a number of different crops and grinding grain. I do give them any surplus milk we have that doesn't go the pigs, and they get high-quality kitchen scraps (equal to Grade I foods from the section on Supplementing Commercial Feed for pigs). I don't feed them in any set proportions, but I experiment trying to find the best combination of foodstuffs I have available. Watch what scraps you feed as they are more likely to affect a day-to-day product such as eggs than a more cumulative one such as pork. In other words, if you fed your chickens a dish of garlic, you would hardly want to have a conversation with one of them in a closed room, much less eat their eggs. I have found a combination of scraps, surplus milk, and ranging cuts our feed bill to practically nothing.

Whenever you are supplementing or substituting for commercial feed, be certain that the flock gets calcium and grit (both found in commercial feed) from other sources. The calcium, available from ground eggshells or ground oyster shells, helps form the eggshells. The grit, picked up naturally by free-ranging birds, enables the toothless chicken to grind its food. For birds not raised free range, a calcium ration and grit should be available if commercial mash is not used for feed.

MANAGEMENT

Routines

The routine a person will want to develop in managing his or her flock will take time, and some mistakes will be made before it suits the family's needs.

A hen will begin laying walnut-sized eggs anytime from $4^1/2$ to 6 months of age. Soon the eggs she lays will be normal size. As my young chickens reach maturity (also signaled by some rather discordant crowing from the young roosters), I usually butcher off all but two of my roosters, keeping a spare in case of tragedy, and keeping the *best* for breeding, and resisting the temptation of imagining how much better they would look on the dinner table.

For a hen's first year you can expect an *average* of an egg a day during her productive period (see Molting section). Depending upon the breed, she will lay from 200 to 240 eggs during her first year and decreasing amounts, though larger eggs, in succeeding years.

You will have to decide whether to keep your hens for more than one year. After the first year, when production stops, the question is whether to keep these at a less productive feed-to-egg-conversion ratio or spend the money to buy and raise new chicks to laying age. This will depend largely on the price of new chicks in your area and the amount of feed you need to purchase or whether you can supply it through your own devices.

If you wish to keep your hens another year you will have to accept the molt with understanding . . . and no eggs. I don't enjoy buying eggs, so I have worked out a routine that enables me to get around molting economically. We started our flock with pullets and after a summer of production, a molt, and a practically eggless winter we hatched out 25 chicks the following spring. By the time the original birds went into molt again the next fall, the chicks had begun laying. We butchered the older birds and took the new layers through the winter (allowing 14 hours of light and plenty of thawed water so as to not upset production) without a molt and hatched another batch of chicks that spring. Again, we butchered the older hens that fall as the new chicks began laying. In this way we never had our hens go through molt, had plenty of meat, and never went through gaps in egg production.

Molting

At the end of a laying year (or at times due to disease or mismanagement), hens molt (or drop their feathers) in order to renew them (Fig. 1.6). They look sickly, *don't lay* (some do but these are rare), and to top it all off, tend to eat more. Depending on the bird, the molt cycle (dropping and growing feathers) can take from 10 to 24 weeks, with little or no egg production during that time. Those taking longer to molt should be culled from the flock. Fast molters often lay for a few weeks after dropping feathers, so egg loss is at a minimum. They tend to drop feathers in clumps rather than singly and grow them back in clumps, hence returning to production more quickly. Slow molters drop one feather at a time and renew them one at a time. They often begin molting earlier and continue for months. Eat those birds.

Fig. 1.6: Wings during different stages of molt. (1) shows the 10 old primary feathers (black), and the secondary feathers (broken outline), separated by the axial feather (x). (2) shows a slow molter at six weeks of molt, with one fully grown primary and feathers 2, 3, and 4 developing at two-week intervals. In contrast, (3), a fast molter, has all new feathers. Feathers 1 to 3 were dropped first (now fully developed); feathers 4 to 7 were dropped next (now four weeks old); and feathers 8 to 10 were dropped last (now two weeks old). Two weeks later (4), feathers 1 to 7 fully grown. Fast molt took 10 weeks, compared to 24 weeks for slow molt. See text for further explanation. (South Dakota Extension drawing.)

When molting begins, feathers are shed in the following order: head, neck, breast, body, wings, and tail. Some birds will molt after only eight or nine months of production (early molters—cull them), while others will lay for 12 months before the onset of molt. The primary feathers of the wing (see Fig. 1.6) are used to chart the molt because they are dropped and renewed in a set order. By examining the primary feathers you can estimate how long a bird has been in molt and how soon before she will return to laying. If the feathers are dropped singly or in groups, you can decide if the bird is a slow or fast molter. The first primary feather to be dropped is in the middle of the wing (Number 1 in the upper left corner of Fig. 1.5) and it takes 6 weeks for a new primary to be fully regrown. Normally, the next primary drops in 2 weeks; the next, 2 weeks after that, and so on. When a fast molter drops 2 or more primaries at a time, they will also grow out together. By dropping primaries in groups of 2 or more, these fast molters (and high producers) complete a molt more rapidly.

Force Molting

For birds that do not normally go into molt after 12 months, or for those birds whose production lags prior to molting, it may be desirable to use force molting (Table 1.6). Through this procedure the birds will drop their feathers and will return into production after new feathers are grown, which may be as soon as 8 weeks. Egg production, as it normally would be in the second year, is lower after force molting. In addition, the eggs are larger, feed consumption is higher, and the health of the flock will be poorer. I have never used it because unlike a commercial poultryman, I do not live or die with relatively minute changes in production, and I also consider it an unnatural strain on a bird. It is, however, an alternative to buying a new set of birds after the first year of production.

Table 1.6: Force Molting Procedure

(A) For Normal Weather Conditions

Day	1	Decrease light to 8 hours per day in light-tight houses or no-natural-light in window houses. Remove all feed and water.
Day	3	Provide water.
Day	8	Provide growing mash—40% normal consumption level.
Day	22	Restore lights to premolt level and full-feed laying ration.

(B) For Hot Weather

Day	1	Decrease light as above. Remove feed but allow water free choice.
Day	8	Provide growing mash—40% normal consumption level.
Day	22	Restore lights to premolt level and full-feed laying ration.

Source: *Raising Poultry the Modern Way* by Leonard Mercia. Charlotte, VT: Garden Way Publishing Co., 1975.

Culling

Culling is the removal of unhealthy or poor producers from a flock. Unless diseased, a cull bird offers the compensation that it is at least as good for the table (maybe meatier due to lack of strain because of egg producing) as a layer. Eat well.

Egg Production

Maximum egg production cannot be maintained without thawed water in the winter or 14 hours of light per day (Table 1.7). As the days grow shorter, artificial light must be supplied so that the 14-hour requirement is met. Lights can be turned on before dawn or after dusk or in combination. Since I will not buy an electric timer to control my lights (it costs too many dozen eggs), I have found that putting the light on at dusk and turning it off before bedtime is most convenient for me.

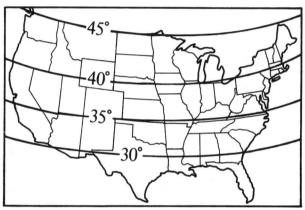

Table 1.7: Hours of Daylight by Month

Month	45° N. Lat.	40° N. Lat.	35° N. Lat.	30° N. Lat.
January	9:09	9:39	10:04	10:25
February	10:26	10:41	10:56	11:09
March	11:53	11:54	11:57	11:58
April	13:29	13:15	13:04	12:53
May	14:51	14:23	13:59	13:39
June	15:35	15:00	14:30	14:04
July	15:07	14:34	14:07	13:44
August	14:06	13:46	13:29	13:14
September	12:33	12:30	12:26	12:22
October	11:01	11:12	11:20	11:28
November	9:34	9:59	10:21	10:41
December	8:48	9:22	9:50	10:13

You must supplement daylight to maintain 14 hours of light per day to obtain maximum egg production. Refer to the map and the chart to determine how many hours of supplementary light (if any) you must supply. More accurate readings may be gained through interpolation.

Allowing laying chickens to range freely over the premises may be delightful to them and help you to save on feed, but it can wreak havoc with gathering eggs. If not confined, chickens have a habit of finding secluded spots to lay their daily eggs. Often the owner is unaware of these sheltered areas. It is a frustrating experience to find a large batch of eggs in a corner of the barn and try to determine their ages. Man has yet to develop a chicken that lays a date-coded egg. Pick up eggs at least once a day and twice a day in cold or hot weather. This serves another function: it helps to avoid the problem of hens eating their own eggs.

Often this urge to lay eggs in unexpected places is due to a filthy nest box, but more often it is the capriciousness of the birds. If I can find out where they're laying (they love haystacks and similar out-of-the-way places), and I can collect them easily, I usually don't bother the hens. If not, I try confining them to the coop for a few days so they have no choice.

Caponizing

Caponizing, or castrating the male chicken either surgically or chemically, is beyond the scope of this book. Leonard Mercia's book, *Raising Poultry The Modern Way*, goes into considerable detail about this subject. If you want to raise them—and they make good eating—capons can be bought from most local poultry farms. They cost about four to five times as much as day-old chicks, and they are sold at about 4 to 6 weeks.

Handling

The best way to carry a chicken is to hold it upside down by the feet. The blood rushing to its head will subdue it after a few moments of indignation. Unless you lock up your chickens at night and can corner them in the coop, catching a chicken is another story. I usually wait for them to put themselves to bed at night or grab one before letting them out in the morning. I have, inadvertently, caught one in a Havahart trap when I was trying to catch rats. Poultry hooks that snag a leg, available from most farm catalogs, make what is potentially a day-long task a snap.

Predators

Many predators can threaten your flock, both during the daytime and at night.

Rats and mice are often the greatest nuisance. They love chicken feed, and will eat it as well as destroy the bags it is kept in. There are many methods of getting rid of them. Keep the feed in ratproof containers. Eliminate the hiding places they prefer, such as the spaces between double-sheathed walls or under floors, or in stored equipment. Set traps for them. Keep a cat. If all else fails, put out anti-coagulant bait. To avoid environmental damage or accidental poisoning of other animals, use a proper bait box that only rats and mice can enter.

Dogs will raid a henhouse or a flock outside, killing some and causing many of the rest to pile up and suffocate. It's difficult to break a dog of this, to them, sport, particularly if it's a neighbor's dog. Try a person-to-person appeal to have the dog tied up before you explore legal options.

Cats are troublesome only with small chicks. Keep the chicks well protected.

Hawks will strike in the daytime, carrying off young birds or eating larger ones on the ground. Owls, too, will use much the same approach. A yard enclosed with a high fence offers some protection.

Many wild animals will dine on chicken, given the opportunity, even eating in a henhouse to do it if you offer an opening. These include foxes, skunks, minks, weasels, and raccoons.

BREEDING

While you can get eggs with just a flock of hens, to get fertile eggs you need a rooster. Hatching your own eggs enables you to raise replacements and cut your ties to hatcheries. One rooster can handle up to 15 hens. (Note: 14 hours of light a day are also essential for maximum fertility of your rooster.) As a rooster is routinely pecking about the yard he will, as if a button on a distant control board was pushed, suddenly look up, begin chasing a hen, collar her by the neck, drive her face into the ground, and mount her. It is hardly the tenderest of nature's mating rituals.

Artificial Incubation

Incubators can be purchased from feed stores and ordered from farm catalogs, but unless you plan to hatch eggs on a large scale, I doubt the economic value of such an operation (and that's presumably why you'd choose to hatch eggs in the first place). It is possible for anyone to construct an incubator with a minimum of time and money (see Appendix).

If you are collecting fertile eggs day by day, you must keep them under carefully controlled conditions until you incubate them. Storage of up to a week has a minimal effect on an egg's hatchability, but after 12 days the viability declines quickly. Eggs must be kept at 55°–65° at a humidity of 70–80 percent. Under these conditions the egg cell remains dormant. At 80°–90° the cell germinates but soon dies; at temperature above 110° the cell will also die. Eggs should be stored, small end down, at an angle of 35° and rotated twice each day. Discard any eggs that are misshapen, cracked, or unusually large or small.

An incubator must be set at a temperature within the very strict limits of 99°–105°. Start the incubator the day before to check its operation and to allow it to heat up before inserting the eggs. Warm the eggs up to room temperature before placing them in the incubator so their temperature will not fall sharply when they are put in. Have a small pan of water on the floor of the incubator to prevent the air from becoming too dry. Finally, turn the eggs 3–4 times a day. Make a small mark on one side of each egg to avoid confusion when turning.

If you have done everything correctly, in 21 days (give a day or two either way) you should notice cracks appearing on the eggs and perhaps hear little peeps coming from them. The breaking-out process is a long and arduous one for the chick and can take up to a day. If after that time a chick hasn't emerged from the egg, it is generally recommended that you leave it and let nature take its course.

However, I can never bear to see a beak poking out and leave the chick to die, so I usually help the little fellow out. Thus far I have never lost any chicks this way, and my henhouse has not been struck by lightning from above.

In planning the ultimate size of your flock, figure on an average hatch of 60 percent. I usually raise my chickens for both eggs and meat so I am not unduly concerned with the sex of the chicks. I just butcher off most of the cockerels when they reach the size I choose to use them.

Once a chick hatches, keep it in the incubator until it is dry and fluffy, about 12 hours. From then on, manage as you would a day-old chick that you've purchased.

Natural Incubation

I find this method much easier because the mother hen will do the brooding and you don't have to build an incubator. On the other hand, a hen will not hatch eggs whenever you want her to. This method is less efficient, too, because it puts a bird out of production for the hatching and brooding period.

All breeds are not equally suited for mothering. Some of the breeds have had the brood characteristic bred out of them. Bantams, Wyandottes, Plymouth Rocks, and Rhode Island Reds make good mothers and will generally set, especially as they get older. If you want chicks, keep an eye out for a broody hen and place some eggs under her if she doesn't have enough of her own. A large hen can cover up to 15 eggs, and a medium one 9 or 10. Make sure she isn't disturbed and have water and feed nearby for her infrequent feedings.

I leave my setting hen in the coop, rather than moving her, as unfamiliar surroundings may break a hen out of her broodiness. If you notice other hens crowding her out of the box and laying too many eggs with her, you probably don't have enough nest boxes, or the others are soiled. Once she sets, she will take over from there. She turns the eggs every day, and when they hatch, she will brood them until the chicks are old enough to go out on their own. Often there is a problem of acceptance of the mother and her clutch by the other birds, but normally after a few weeks the new family will become accepted. At times we've had chickens become wild, roosting in trees and retuning to the coop only for food. In this case, lock them in the coop for a week and these chickens should become as domesticated as the rest of the flock.

HEALTH

In combatting diseases, your most effective tack is prevention. Letting your chickens range free to reap the benefits of fresh air and sunshine is one way to promote good health. Disease in my flock, and most small flocks, is not a major problem. Disease is more likely to occur in confined commercial flocks, which are managed in a much more intensive way. Signs of ailing birds are a drop in feed and water consumption, lower egg production, and listlessness. Unless you are going into large-scale production, the purchase of a veterinary text devoted to chickens is unnecessary. The disease table in the Appendix will help.

An excellent source of information relating to poultry diseases is the department of veterinary science at your state college of agriculture. When you need advice or treatment recommendations, a call to this source will result in a wealth of information. If you have unexpected losses, presenting a dead bird will often provide the information the poultry pathologists need to answer your questions.

BUTCHERING

Chickens are excellent animals for starting your livestock butchering operation as they are relatively easy to butcher (Table 1.8).

Table 1.8: Butchering Terminology for Chickens

Term	Sex	Weight or Age	
Broiler	Either	2^1/$_2$ lbs.	Less than
Fryer	Either	2^1/$_2$–3^1/$_2$ lbs.	8 months
Roaster	Either	3^1/$_2$–5 lbs.	old
Capon	Castrated male	6–8 lbs.	
Fowl	Female	Over 8 months	
Cock or stag	Male	Over 8 months	

When you decide to butcher a bird, you should deprive it of food, but not water, for 12 hours.

When catching a hen for butchering, watch your roosters. They are never pleased to lose a member of their harem and can get quite nasty. Our rooster clucks indignantly as I stalk one of his lovelies and then attacks me after the catch, but only after I turn my back. I can usually outrun him.

Whatever your choice of butchering size and age (and whatever the personality of your rooster), you will have two methods of killing. A third, breaking the neck, is rarely used. The most popular are beheading and sticking and cutting the throat. Both are equally effective.

Perhaps beheading is best for the beginner since it is the quickest and surest. Holding the bird by the legs, I tie some bailing twine around its feet so that it can hang for bleeding after I've cut its head off. (By hanging it this way you will also be spared the anguish of your first chicken flapping madly about the yard without a head.) Have a sharp ax and a chopping block (a stump will do) with two nails driven in and sticking up 2 inches parallel to each other and about 1 inch apart (Fig. 1.7). Insert the chicken's head between the two nails and pull gently. The beak will get caught between the nails and the neck will stretch, giving you a surer shot. After chopping the head off, hang the bird up immediately and let it bleed out completely. I usually let it hang 10 minutes or so; much longer, especially in cold weather, will make the bird harder to pluck.

When sticking, tie the bird's feet and then hang it upside down. Using a sharp, short-bladed knife, cut the throat at the base of the neck, being sure you sever the jugular vein. Immediately insert the point of the knife into the bird's mouth and force it through the roof toward the back of the head into the brain cavity, and give the knife a quarter turn. The piercing of the brain causes a squawk and a convulsive wing flapping that loosens the feathers and makes the plucking easier.

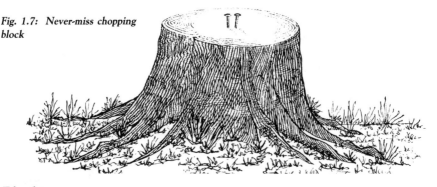

Fig. 1.7: Never-miss chopping block

Plucking

Plucking may be done by the wet or the dry method. Begin dry plucking immediately after the bird is bled and before it cools, since this tends to set the feathers. This is best done after sticking, rather than beheading. For wet plucking, dip the bird in hot (150°–190°F.) water for a half minute to a minute. Too long and the skin will cook. If you wish to freeze your bird for any length of time, it is best to dry pluck since scalding tends to cut down on freezer life.

Plucking is hardly fun, and after doing a few birds, you may be ready to switch to rabbits for your source of meat, and just raise chicken for eggs. Your skill and speed at plucking will improve with practice. It is best to take your time at first so you will not tear the skin and thus expose the meat to drying. The best order of plucking is to do the wings, breast, body, back, legs, and finally the neck. If any small pinfeathers are left, they may be removed with a dull knife. Small feathers and hair may be singed off with a gas flame or a candle.

Dressing

It is best to work with a chilled carcass. A sharp knife is essential. Poultry shears are helpful but not necessary. First cut off the head and, peeling back the skin, sever the neck and save it for the stockpot. Then remove the feet at the hock joints. (After washing, the feet make an excellent addition to the stockpot.) Make a cut in the vent large enough to accommodate the hand. If the bird is of laying age you may find a fully formed egg and see a string of yolks in varying stages of development. Pull out the innards, saving the heart, liver, and gizzard. Carefully remove the gallbladder from the liver and remove the inner sac from the gizzard. Push your fingers in the front and remove the linings or "lights." Flush with water and make sure all pieces of blood clots are removed. After the bird is washed thoroughly, it may be roasted, cut up for frying or broiling, or frozen.

Ducks

Ducks are hardier than chickens. They require less attention and they grow more quickly. While they do not forage for food as much as geese, they will still augment their diet by ranging, and they will produce tasty meat very economically in only 8 to 10 weeks.

Ducks need very little space and almost no housing. Your ducks will not require a pond or a brook, but if you have one, they will be happy birds. The meat is delicious, not at all gamy, and with less fat than wild duck. You also have the opportunity to gather duck eggs and even some duck down for pillows or whatever else you may choose to do with it.

Ducks have something else over chickens. They have personalities! Few people get pleasure from watching chickens, as they have the maddening tendency to behave the same way all the time. Ducks frolicking on a pond or waddling mechanically in a line to some distant object are a sight to behold. Be careful: you can make ducks into pets (try that with chickens), but if their ultimate destination is your dinner table, this can put a crimp in your plans.

BREEDS

Some breeds of ducks are well suited to meat production and others are egg producers. Unlike chickens, there is no such thing as a dual-purpose breed of duck. You can eat good egg layers but the carcass will be small and of poorer quality, since most of their energy goes into the production of eggs, not meat. The meat birds will supply you with a trickle of eggs along with meat. Duck eggs, while larger than chicken eggs, are good for baking, but many people find them unsuitable for "straight" eating. If unfamiliar with their taste, try them before you decide to live on them.

Meat Breeds

White Pekin: This is the Long Island duck you've heard about and is the major breed raised commercially in the United States (Fig. 1.8). Pekins are large birds with white plumage, orange-yellow bills, reddish-yellow legs and feet, and yellow skin. Adult drakes (the males) reach a weight of about 9 pounds while adult ducks (the females) reach about 8 pounds. They produce excellent meat and reach market weight of 7 pounds in 8–10 weeks. They are also reasonably good egg producers, a duck laying an average of 160 eggs per year. They are high-strung birds and, as a result, are poor setters. If you wish to hatch their eggs, it must be done by artificial incubation or by enlisting the services of a broody hen.

Muscovy: While many varieties of this native of South America exist, the white ones, because they dress out more attractively, are most desirable for meat production (Fig. 1.8). They have white skin and the darkening around the eyes resembles a mask. Adult drake weigh 10 pounds and ducks weigh about 7 pounds. They reach market weight in about 10 to 17 weeks. They are poor egg producers (average 40 to 45 eggs per year) but they are the best setters of all meat producers.

Aylesbury: This is the English counterpart of our White Pekin and its quack has a characteristic British air. Like the Pekin, they reach meat weight (7 pounds) in 8–10 weeks. They have white feathers, white skin, flesh-colored bills, and light orange legs and feet. Adult drakes weigh 9 pounds, ducks 8 pounds. While they are not as nervous as Pekins, they do not have much interest in setting. They are not quite so prolific in egg production as Pekins.

Other meat breeds include Rouen, Cayuga, Call, and Swedish.

Fig. 1.8: Examples of duck breeds

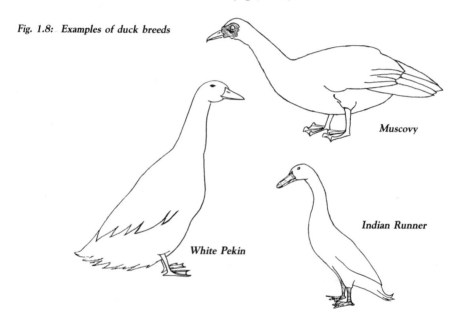

Muscovy

Indian Runner

White Pekin

Egg Breeds

Khaki Campbell: Khakis are able to surpass even the highest producing chickens in egg production. They are, however, not valued for their meat since drakes and ducks reach a mature weight of only 4¹/₂ pounds.

The Khaki drakes are bronze on the lower back, tail coverts, head, and neck; the rest of the bird is khaki-colored. Their bills are green and they have orange legs and feet. The females have seal-brown heads and necks; the rest is khaki-colored. Their bills are greenish-black and their legs and feet are brown.

Indian Runner: Indian Runners fall short of Khakis in egg production, but they are good producers (Fig. 1.8). There are three types: White Penciled, Fawn, and White. All three have orange to reddish-orange feet and legs. They stand quite erect, their carriage being almost perpendicular to the ground. Their adult weight is about that of Khakis. They are not valued for their meat.

PURCHASE

As with chickens, you have three choices: hatching eggs, buying day-old ducklings, or buying older stock for breeding. Your choice will depend on how you will raise your birds. If you plan to have only a few ducks for meat, buy a few day-old ducklings from a local hatchery or friend and raise them. This frees you from getting incubating equipment and fussing with breeding and hatching. If a larger flock is desired, you can buy fertile duck eggs or buy a lot of ducklings and go from there, selecting the best for future breeding stock.

Hatching

The hatching time for duck eggs is 28 days for all breeds except Muscovy which is 35 days. You can use the same incubating and brooding methods as with hatching chickens. The eggs should be placed in the incubator small end down, or if that is impossible, rest them on their sides.

For natural hatching, Muscovys are the best breed to use and frequently the only ducks that will hatch eggs. If you don't have a setting duck, you can use a broody hen. A hen will handle up to 10 duck eggs (don't let her take more than she can handle since this will affect the hatchability of all of them); a duck, slightly more. Sprinkle warm water on the eggs each day if you are using a hen or a duck that has no access to water since duck eggs need more moisture than chicken eggs do. When they hatch, in the case of a duck mother, let nature take its course; in the case of a hen foster mother, confine her and her brood of ugly chicklings so that she cannot wander off as they will not be able to keep up with her the way chicks do, and they may get lost. After 4 weeks the ducklings should be able to keep up with her and she can be set free.

Day-Old Ducklings

In purchasing day-old ducklings locally, you have the advantage of being able to buy fewer (mail-order minimums are often 25), but you may not be able to have your choice of breed. In any event, you will need a brooder. Handle brooding as with chicks but allow 1 square foot of space per bird. Ducklings normally need brooding for only 4 weeks and may even get by with 2 or 3 weeks. In warmer summer weather a light bulb in their shelter may suffice. Do not allow a duckling to get soaked by rain before 4 weeks of age or until it is feathered. It should also not be allowed to swim before 6 weeks because the oil is not fully distributed on its feathers, and it will be more susceptible to chills and sickness. Shade must also be available to young birds.

Older Stock

This applies to choice of mature stock which are bought for breeding or to picking potential breeders from your own flock. At 6–7 weeks drakes and ducks can be distinguished not only by the difference in coloration (males have brighter plumage), but also by their sounds. Females "honk" and drakes "belch." You will require one drake for every 6 females for breeding. Look for ducks with evenly colored feet, legs, and bills. Choose birds that are vigorous, alert, heavy, and solid, with broad breasts and necks that are not too long. Plumage should be even and glossy. Find out, if you can, the breeding records of the parents; choose ducks from parents that have exhibited high fertility, hatchability, and egg production.

Predators

Predators are generally the same as for chickens. Your dog, while perhaps accustomed to chickens, may see the newcomers as fair game. Watch him.

HOUSING

If you plan to raise ducks for meat in the summer months, housing is quite simple. If you don't have a handy pond or stream, you can house them with the chickens—if they will stand for it. If the chickens peck and chase the ducks, as chickens may do, make a little pen within the coop that only the ducklings can scoot under. Within a few weeks they will probably be accepted, but if not, they will be close to butchering age anyway and out of your and your chickens' hair (or feathers).

Allow ducks to range free all day, then lure them with grain into a small enclosure and lock them in at night to safeguard them from predators and hungry neighbors. A simple, movable hutch should have a dry floor and furnish protection from severe weather and predators (Fig. 1.9). They don't need roosts, but allow 5–6 square feet of floor space per bird.

Fig. 1.9: A movable house for ducks or geese

If you wish to raise ducks, you will need a more permanent house. It can be much the same design as a chicken coop, but no roosts are needed. Ducks are more resistant to cold than chickens, so you don't need to be so careful about plugging up drafts in the building. However, you shouldn't allow any snow or rain to seep in, and the house should, of course, be predator-proof. Space requirements are the same as for the temporary shelter described above. If feather pulling is a problem, the ducks may be overcrowded. Give them more space. If this doesn't work, debill them (clip off the forward edge of the upper bill). Keep the house scrupulously clean. A built-up system works well. Be sure to shovel out any wet spots.

If you allow your birds to range, as I strongly suggest, they will be healthy and disease-resistant. Keep them away from pools of stagnant water, since these are havens for disease. If for any number of reasons (limited space, predators, large flocks) you must fence your birds in order to let them run, allow at least 40 square feet per bird. Keep the run very clean. Because ducks can be quite messy, sand is the best ground cover and it helps if you locate the run so that it slants downhill away from the house. For run-confined birds, you must throw in fresh greens daily to realize any feed economy.

EQUIPMENT

Ducks do not need swimming water, nor is it necessary for fertilized eggs, but if you have a pond, your ducks will love you for it. Don't allow the population to get too high or your pond will be a mess and will smell awful. For ducks confined to a run or on a pondless farm, supply them with a pool they can splash in and clean themselves. This is especially good if you have ducks hatching eggs, since it will enable the duck to supply more moisture to the eggs. The pool should measure at least 3 feet by 1 foot and be deep enough so the ducks can totally immerse themselves. It can be constructed out of concrete or wood, or you can use an old feed trough or discarded bathtub. Change the water frequently to prevent stagnation.

Feed Dishes

Your ducklings can be fed from chick feeders or shallow troughs or pans. Since they do not scratch their food as chickens do, waste is less of a problem. As they get older, you can feed them from larger troughs or from a self-feeding hopper (Fig. 1.10). This hopper is very useful if your birds are free-ranging since you don't need a larger house to hold the feeding equipment. If you have free-ranging chickens, food piracy may be a problem. We have handled this by separating them from each other's ranging area.

Waterers

The duck's waterer must be constructed so that the whole bill (but not the whole body) can be immersed and the nostrils can be cleaned. Ducks are very messy drinkers, so if they are watered inside their house, place the waterer on wire

3′–4′

off the floor and have a lot of absorbent litter underneath. Better yet, don't give them their water while they're cooped up. They can adjust to being without it while closed in at night and their egg production and growth will not be affected. If you keep the waterer outside, the food must be kept out of the coop, too, since a duck eating dry food without access to water may strangle.

Nest Boxes

If you keep ducks for egg production or breeding, nest boxes must be provided. They can be constructed like chickens' nest boxes, but should not have tops. They should be a bit larger, at least 15 inches square and 18 inches deep, and they should be set on the floor. Nest material can be the same as for chickens.

Stampede Lights

When startled at night, ducks have a tendency to run in circles, possibly causing injury or death. A low-wattage night light will keep this from happening.

FEED

Ducks benefit (or maybe it should be said that *you* benefit) from an excellent feed conversion ratio (3:1), and this should enable you to raise meat and eggs economically. While they are not as good at foraging for food as geese, ducks will augment their diet if they are allowed free range, or supplied with greens if penned.

Table 1.9 outlines the rations that should be fed to ducks:

Table 1.9: Feeding Rations for Ducks

Starter for the first 2 weeks (20%–22% protein)

Grower for meat birds, 2 weeks to market (15%–18% protein)

Breeder developer for breeding and egg flock, 6–7 weeks of age until 1 month before production

Breeder from 1 month prior to production

Pelleted feed, if available, is preferred to mashes, as there is less waste and the ducks convert it to meat at a better rate. Most feed stores should carry duck feed or feed that is suitable for them. Chicken feed can be substituted for *growing* flocks, but *medications present in chick starter can kill young ducklings*, which should be fed either an unmedicated chick starter or feed formulated for ducklings. The starter feed can be in the form of pellets (no larger than $1/8$-inch) or mash that is slightly wet. Allow the feed to be consumed free choice and supply fresh water.

For meat ducks, allow free choice all the grower feed they will eat. Again, mash can be used but a $3/16$-inch pellet will be utilized with greater efficiency and less waste. The ducks you plan to use for breeding should be fed the breeder developer ration. To prevent them from becoming too fat, feed this on a limited basis— $1/2$ pound per duck per day, split between morning and evening feedings.

At one month prior to egg production (explained in section on *Breeding*), switch the ducks from the breeder developer ration to a breeder ration and feed them free choice. If feed does not contain a calcium supplement (which is unlikely if it's a commercial feed), ground eggshells, chicken or duck, should be added.

Supplementing Commercial Feed

While their feed conversion ability is good, and their relatively small level of feed consumption makes supplemental feed less important than for pigs, you can still make some savings. It is impractical to mix your own feed, but you can augment their diet by allowing them to free-range. They will obtain a great deal of their feed requirements naturally and balance their diet at the same time. Ducks that are penned should be supplied with *fresh* greens at least once daily.

BREEDING

Ducks that have hatched from April through July will reach maturity at 7 months. Those hatched between September and January will mature in 5 or 6 months because of the increasing length of the days in their growing period. It is best to pick your future breeding stock when they are 6–7 weeks of age and begin feeding them as indicated in the *Feed* section. (The sexing of ducks is discussed in the *Purchase* section and under *Breeding* in Part III, *Geese*, that follows.) You will need one drake for every 6 ducks, but I suggest at least one extra in the event that one of the drakes meets an untimely end.

When your birds are near maturity, and 3 weeks before you wish them to begin full egg production, increase the light in their coop so that they receive 14 hours of light a day. Give the drakes an extra one or two weeks' head start so you'll be sure they're ready for the ducks. Bringing females into full production (supplying them with 14 hours of light) is not recommended before 7 months of age because they will tend to produce smaller eggs with a lower hatchability.

Once the duck begins laying, she will reach her production peak within 5–6 weeks. If you are interested in hatching eggs, fertility is highest when egg production is highest, and the greatest hatchability will occur after the first few settings, probably because the birds will have improved mating proficiency. The details of handling fertile eggs, incubation, and brooding are given in earlier sections.

Ducks lay most of their eggs before 7 A.M., so keeping them locked up until after they are through laying will not markedly affect their feed consumption (if you keep the feed outside) or their ability to range. The number of eggs they lay will, of course, depend upon the breed. High production will be maintained for 5–6 months and then will taper off gradually, allowing the ducks a rest period.

HEALTH

Prevention is the byword. Clean pens and runs frequently and avoid crowding and you shouldn't even need to consult the disease chart in the Appendix. Keep young ducks from becoming chilled and keep all ducks away from stagnant water.

BUTCHERING

Ducks reach market weight at 7–10 weeks of age (except Muscovy: 10 to 17 weeks). Butcher those with firm, plump carcasses without many pinfeathers. Raising them beyond the recommended ages and weights will perhaps produce larger ducks, but not so economically.

Keep ducks from food for 12 hours before slaughter but allow access to water. Methods of killing are the same as for chickens. You can dry-pluck or scald; wax on the pinfeathers may be quite helpful.

 Geese

Geese are the hardiest of all poultry; they are the most disease-resistant and the cheapest to raise. Because they are such good foragers and can be set out to forage when they are young, they can be raised with an absolute minimum of feed. They also make good "watchdogs," a plus in crime-ridden suburbs. Walk a goose every night, and see if you get mugged. They are good weeders, their feathers and down can be sold for extra cash, their meat is excellent (and abundant), and don't forget *pâté de foie gras!*

And, yes, they are very entertaining. To me, they are the reincarnation of an especially "rammy" ram. People we know have geese that were especially playful and, in fact, enjoyed playing with their children. One day as friends were visiting and they were riding their snowmobile, one of the geese jumped on the back for a ride. Naturally, the friend, upon seeing this, jumped off. Away the goose rode, honking madly, across the field and into a ditch.

BREEDS

There are breeds of geese to suit your various tastes: meat, eggs, eggs and meat, show, and pets. Table 1.10 gives the major breeds of geese and their size, as recommended by the American Poultry Association in the Standard of Perfection.

Table 1.10: Breeds of Geese and Their Weights in Pounds

Breed	Male		Female	
	Young	*Adult*	*Young*	*Adult*
African	16	20	14	18
Buff	16	18	14	16
Canada	10	12	8	10
Chinese	10	12	8	10
Egyptian	5	5$^1/_2$	4	4$^1/_2$
Emden	20	26	16	20
Pilgrim	12	14	10	13
Sebastopol	12	14	10	12
Toulouse	20	26	16	20

African: The African has an erect body that is carried high from the ground. Its head is light brown with a large black knob atop. It has large dark brown eyes and a black bill. The feathers are light ash brown on the underside of the body and on the neck and breast, and a darker ash brown on the wings and neck. African geese are good layers and they grow rapidly to market weight, but they have more pinfeathers than most meat birds. They also tend to be very noisy.

Buff: This breed has poor meat and egg-laying qualities and is not worth considering for our purposes.

Canada (wild): These geese are not good meat or egg birds and thus are valuable only for pets or show. You need a permit to own one. A Canada goose can be kept confined only in a wire pen or by clipping their wings. They are better off left in the wild.

Chinese (brown or white): This is a good multipurpose breed. Its graceful, swanlike appearance makes it popular for show or pets; its relatively high egg production (40 to 65 eggs per year) makes it a desirable egg bird (Fig. 1.11). It is of medium weight, grows rapidly, matures early, and its meat is delicious. As with all poultry, the white variety is more desirable for meat purposes because of the absence of dark pinfeathers. Chinese geese also make very good "watch-dogs."

Egyptian: Due to small size, worthwhile only for pets or show.

Emden: A pure white, erect goose (Fig. 1.11). It is a reasonably good layer (35 to 40 eggs per year) and a good setter. It is also a good meat bird as it grows rapidly and mature early.

Fig. 1.11: Examples of goose breeds

Pilgrim

Chinese

Emden

Pilgrim: The Pilgrims are good market birds, and they have the added distinction of being sex-linked (color-coded by sex), a decided advantage for those of us who are not terribly proficient at feeling a goose's sex organs to determine sex. The day-old male goslings are creamy white and the females gray. The mature ganders (male) are white with blue eyes; the geese (females) gray and white with dark hazel eyes.

Sabastopol: This lovely white goose with curly feathers is good only for show or pets.

Toulouse: Toulouse geese make good mothers and lay about 20 to 30 eggs per year. They are also quite good as meat birds. They have broad bodies and look even larger because of the fluffiness of their feathers. They are dark gray on the back and the breast and white on the abdomen. Their eyes are dark brown to hazel, the legs and toes are deep reddish-orange, and the bill, orange.

Geese can benefit from crossbreeding. A particularly good cross is a White Chinese gander with an Emden goose. The result is a rapidly growing market bird. Experiment!

PURCHASE

You can buy fertile eggs for hatching, day-old goslings, or mature stock for breeding. If you plan to go into breeding, get sexed birds or buy enough so that you're sure you have a gander (preferably two, in case of accidental loss), and you won't have to buy one when your geese are grown. Flocks of geese tend to be very tight-knit, so if you plan to introduce a new member, prepare yourself for a battle.

Hatching

In an incubator the procedure is the same as for chickens, except that the hatching time for Canada and Egyptian geese is 35 days and 29 to 31 for the other breeds. Incubators are difficult to use because more humidity is needed than for hatching chickens. Have a pan of water on the floor of the incubator, and try dipping each egg in lukewarm water for half a minute each day. Hatchability begins to decrease after the first week, but eggs can be stored up to 2 weeks before incubation without losing complete viability.

A goose or a duck foster-mother hatcher (preferably a Muscovy) can handle 10 to 12 eggs. Broody hens must be free from lice, and because of the size of the eggs, an average hen can handle only 4–6 goose eggs. The eggs need not be sprinkled with water if the goose or duck doing the hatching has daily access to swimming water. In the case of chickens, or ducks or geese without access to water, dip the eggs in water daily as in incubation. Eggs must be turned daily as they are too big for a hen to turn herself. Also, when a hen is used, goslings must be removed from her as they emerge and returned when all of them have hatched. If she deserts them or wanders away, confine her with them.

Day-Old Goslings

The brooder can be the same as for chickens but allow ¹/₂ square foot per gosling at the start, and increase to 1 foot at the end of two weeks. Depending upon the weather, artificial brooding is needed for only a couple of days to 2 weeks. If the weather is warm and sunny, you can let the goslings out after a few days, but be very sure they don't get caught in a shower or exposed to heavy dew. If you don't have a chicken brooder, you don't need to build one as elaborate as for chickens. Because they are so hardy, goslings can be brooded in a box with a light or in the corner of a room or barn. Make sure that no predators such as rats or weasels can get to them.

HOUSING

In most cases, housing for geese is minimal or none at all. Geese raised in warmer months need a simple shelter so that they can be confined at night safe from predators; if predators are not a concern, they may be left to find their own shelter. A more permanent shelter for breeding purposes can be the same as indicated for ducks. Allow at least 8 square feet per goose for the larger breeds. Don't worry about them in winter as they love to frolic in the snow and may spend more time outside the pen than in it.

Farms having a large pond that does not freeze completely can allow geese to remain outside all winter. Geese have a innate sense that makes them remain in open water during the night, thus minimizing the chance of a predator attack. Snapping turtles are hibernating at this time of year so pose no threat.

Don't just plop your geese down anywhere. Geese will roam quite a bit, so it is a good idea to house them as far from your garden as feasible. They can be quite messy with their droppings and graze very closely, so rotation of pastures (if fenced in) or a location out of barefooted humans' traffic patterns is recommended. If you are forced to fence your geese because of predators or angry neighbors, use a 3-foot fence. Geese will also tear the bark from young trees, so keep them from young fruit trees, or other trees you don't want to lose, or wrap the trees with wire as high up as the geese can reach.

EQUIPMENT

Feeding and watering equipment is the same as for ducks. The range feeder (a trough in which feeds in which scratch or mash can be placed) is especially important for large groups of geese. The waterer should be deeper so that they can immerse their heads, but not so big that they can swim in it. Swimming troughs, like those for ducks, can be provided if there is no natural water. Do not let them swim before 3–4 weeks of age or chilling may result.

Your nest boxes can be located inside or outside the house because more often than not, geese like to lay their eggs outdoors. If the boxes are outside, place them in some sort of natural shelter or build a rudimentary one to protect them from the elements. You will need 1 nest for every 3 birds, and it should measure 2 feet square. If possible, space the nests a distance apart from each other to discourage fighting.

FEED

Fresh water must be available at all times.

Geese should not be fed medicated chick starter. Try to find a duck or goose starter (20%–22% protein) and feed that for the first 3 weeks. If you can't find any starter, feed them bread crumbs wet with milk or a mixture of half cornmeal and half bran or oats moistened with milk. If the weather is suitable, you can let them out to forage for greens. In inclement weather, supply them with greens in the brooder (along with their regular feed).

The rest of their feeding program is the same as for ducks, except that you can rely more heavily on feeding whole grains, and geese will forage a great deal more than ducks do. Mix whole grains with the mash or pelleted feed at a 50:50 ratio. If you wish, for greater economy, you can feed only whole grains: corn, oats, bran, or wheat. If you are not feeding commercial feed with grit, you should supply some grit to your flock if they don't get enough in their foraging.

Geese favor the more succulent, tender grasses like clover, eschewing old pasture and hard stems and even alfalfa. Geese can get by on almost no grain the first month (and they don't seem to eat very much anyway) and therefore can be raised for very little cost. If you raise your geese on forage alone, finishing will be important for the best quality meat. If you confine your birds to a run and they exhaust their greens, it is essential that you supply them with plenty each day.

Finishing

A month or 3 weeks before slaughter, gradually work your geese over to a diet of $1/2$ to $1/4$ pasture and the remainder fattening feed. This finishing feed can consist of whole corn, cornmeal, barley, or a cornmeal and mash (or pellet) mixture moistened with milk. What you use depends mostly on what is available and cheap—experiment! As with any animal, do not change their feed overnight but introduce the new feed gradually.

Breeding Diet

If you are wintering geese for breeding (or if you have become attached and want them as pets), you can feed them as for breeding ducks or a diet consisting of 15 to 25 percent breeder feed and the rest good, tender hay (preferably legume—second cutting is best) or silage.

Supplementing Commercial Feed

Their foraging ability makes geese one of the most economical meat producers to raise. In addition, you can grow much of the feed in programs outlined above. You can plant a field of clover to supply exceptional grazing for your flock. Allowing geese to forage and feeding them corn (or barley, oats, etc.) that you grow yourself should reduce feed cost to an absolute minimum: the cost of the seed itself. With very little effort you could grow all the grain you need for half a dozen geese.

Geese have the reputation for being natural weeders. It is my observation that this unique ability has been overrated. I have spent many summer hours waiting for my geese to eat the weeds in my garden, and for some reason I always end up doing the weeding while the geese watch in the shade of a tree.

MANAGEMENT

Handling

Geese will follow you most anywhere if you coax them with a little grain. If you want to pick them up, secure them as gently as possible by the base of the neck, reel them in, and hold the wings gently with the other arm. Grabbing them by the legs, as with chickens, can cause injury.

Predators

Geese are threatened by the same predators as other poultry but are able to ward off the smaller and more timid animals. Our dog, chicken killer of repute, has yet to tangle with a goose. They may not be able to fight off large dogs, but they are convincing bluffers. Still, a good house to lock them in at night, and as a last resort, a pen, will do much to prevent losses.

Goslings are, of course, more prone to predator loss. Rats and weasels can wipe out a flock if they can get at them in the brooder. Make sure the brooder is predator-free.

BREEDING

Select geese for mating that are healthy and rapid growers. One gander will service about 4 geese. They are not necessarily monogamous as some people suggest (except the Canada goose, with which we are not really concerned). Once they establish mating partners, it is difficult to introduce new geese to the flock. Generally, the laying and meat birds tend to mate in pairs or trios (i.e., one gander to 2–3 geese), while the smaller breeds may mate one gander to 4–5 geese. The best practice is to have plenty of ganders and watch their breeding habits to see if any of the ganders are unneeded. If they are, butcher them.

Sexing, in all breeds except the Pilgrim, is not the easiest thing in the world. To determine sex (and this takes practice), place the bird on its back (pointing the tail away from you) and bend the tail over backward a bit. Massage the genital area to relax the muscles and then press around the area with the thumb and index finger to expose the sex organs. The goose's will be rounder and more prominent. Practice on birds whose sex you know, and in time, it will be routine. This also works with ducks.

The Chinese may begin laying in the winter, but most geese will begin to lay in February or March as the hours of daylight begin to lengthen. They can be brought into production at the age of 7 months by increasing their exposure to artificial light. If a good pelleted breeder feed (or mash) is available, begin feeding that a month before production. If no duck or goose breeder is available, chicken-laying mash or pellets can be used. If you are feeding a home-prepared feed, be sure to supply calcium (in the form of ground limestone, oyster shell, etc.) for eggshell hardness. Gathering eggs frequently will help cut down on broodiness. Hatch-ability will increase after the first year of production and is at its peak between 2 and 5 years. Geese will lay until about 10 years of age (and will often live many years past this), and ganders are good breeders for about 5 years.

During the breeding season, geese tend to wander in search of a nesting ground if they are allowed to roam freely. In some settings this characteristic may be undesirable so it is necessary to restrict their range in the early egg-laying season.

Finding goose eggs may present problems for the owner but it never seems a problem for the local raccoon population, which apparently considers goose eggs a treat. It's unlikely that a goose can hide her eggs from the raccoons, so collect the eggs as they are found and keep them in a cool place until a proper brooding area can be constructed. In our experience, allowing geese to set on their eggs outside is doomed to failure.

HEALTH

Geese are among the most disease-resistant of all livestock. Normal preventive measures should insure your flock against losses due to disease. The Appendix lists common poultry diseases, should you have problems.

BUTCHERING

Geese are ready for butchering at about 14 to 16 weeks. You can butcher at 12 weeks if you feed them extra heavily, but then again you're not saving money (and in fact are probably spending more), just time, and who's in a rush? Also, at the earlier age you'll have to do battle with more pinfeathers, and that is certainly no delight. The best time to butcher is in the fall as the geese reach market weight and the supply of forage is dwindling.

Slaughtering is carried out as with other poultry. Save the blood (there's enough of it with geese to make it worthwhile) and the offal for other stock. The feathers are the worst to remove of all the poultry, but scalding and coating with wax, then finishing the job with a blowtorch, should make an attractive carcass. Some people hang the gutted carcass for a few days to age; suit yourself.

Down

There may be some market for the down in your locality, or you can most certainly make use of it yourself for stuffing pillows, quilts, sleeping bags, or your own parkas. If you wish to save the down, don't use wax when plucking. Wash the feathers in lukewarm water and detergent or water and borax and washing soda. Rinse them out and pat dry, then hang in an airy place in a cheesecloth bag.

You can also pluck live *mature* geese each spring and "harvest" the down year after year. Only mature geese should be used, and there will be no pinfeathers. Restrain the goose and pluck the down parallel to the body and toward the tail. Do this only after the weather is settled in the spring and do not pluck them bald. They get embarrassed. Wash and dry the feathers as you would after plucking dead geese.

 Turkeys

"When God gave out brains . . ." the old saying goes. Well, that day the turkeys got tied up in traffic. Ditto when they gave out resistance to diseases. In modern parlance, a "turkey" is a worthless individual, an idiot, a dolt. These comparisons hold, however, only for domestic turkeys. Wild turkeys survive quite well on their own—out of the reach of man. It seems when we domesticate animals, "improve" them for our own purposes, we make them dumber.

Despite their apparent shortcomings, no one seems to find much fault with their taste. Turkey is such a mainstay of American living, that it was included in the first meal served on the moon. But when it comes to the home-raised, freshly killed turkeys, there is nothing—repeat, nothing—like their taste.

Turkeys are not the easiest poultry to raise, but raising them is definitely within the capabilities of the average homesteader. They require little space and time, and with sanitary conditions they should be disease-free. Most anyone who raises animals should consider raising a few turkeys over the summer—a couple for family use, and a couple for sale to pay for the ones they keep and even make a little profit. If ever your turkeys get you down, if their stupidity amazes you and your wish to be rid of them on the spot, be humbled by the old saying: "The only thing stupider than a turkey is the one who raises them."

BREEDS

There are three breeds that you will be most likely to come in contact with: the Broad-breasted Bronze, the Broad-breasted White and the Beltsville White. The Bronze is the breed everyone pictures when they think of turkeys. Their plumage is brown or black, and because of the problems in plucking a dark-feathered bird (getting a clean-looking carcass), the White has pretty well replaced the bronze as a commercial bird. The Beltsvilles are smaller and make good broilers at 15–16 weeks and good medium-size roasters at 24 weeks of age. The larger birds (Whites and Bronzes) make good large roasters at 24–28 weeks. The large-type hens can be butchered at only 13 weeks as a fryer-roaster and can be butchered at 20 weeks to make good medium-size roasters. The main difference between the Beltsvilles and the larger Whites and Bronzes is that the Beltsvilles do not convert feed quite so efficiently, but they are usually cheaper to buy as poults (young turkeys).

Fig. 1.12: Broad-breasted White turkeys

PURCHASE

You will be able to choose between the purchase of eggs and poults. Breeding turkeys is beyond the scope of small operations. Turkey eggs can be incubated much the same way chickens' eggs are, except that the hatching time is 28 days rather than 21. After hatching, or upon purchase of poults, you will have to place the turkeys in a brooder. Brooding is the same as for chickens.

Turkeys must never be housed with chickens, nor should they be kept in the pen or brooder that has housed chickens within the past three months or where chickens have ranged within three years. These precautions are to prevent turkeys from contracting diseases. If you use a brooder or pen before the time has elapsed, disinfect it thoroughly.

Allow the poults twice as much space and headroom as you would for chickens. Turkeys are more sensitive to drafts and dampness, so take pains to eliminate such conditions. Also, cover the litter with rough paper for a week to prevent litter eating. Litter is not needed if you raise them off the floor on wire or wooden slats. Weather permitting, they may be removed from the brooder in 8 weeks. Round corners in brooder pens with curved cardboard or hardware cloth, to prevent piling up in corners and smothering.

Purchase your eggs or poults from reliable breeders who guarantee disease-free stock. While you can order turkeys by mail, they do not ship as well as chickens, so you should try to buy locally.

HOUSING

There are three types of housing: confinement, sunporch, and ranging.

Confinement

This is very popular because it protects the birds from most predators as well as severe weather conditions. Again, dryness is one of the primary concerns. Litter must be deep and dry and more added as it becomes damp. The birds can also be raised off the floor on heavy hardware cloth or on boards that are slatted $3/4$-inch apart. Roosts are not necessary, but if they are desired, construct them as you would for chickens. Allow at least 1 foot of space per bird on the roosts. In confinement housing, allow 5 square feet of space per bird if they are debeaked, as described in the chapter on chickens, and 7–8 square feet if they are not. The drawback of this system and the sunporch method is that the turkeys won't be able to make use of greens and other supplementary feed the way range-reared birds can.

Sunporch Method

This is really identical to the above with the addition of a wire- or slat-floored "sunporch" attached to the house (Fig.1.13).

Ranging

This can be accomplished by letting the birds out of their confinement house and allowing them to free-range. If you have chickens, they *must not* mix. You can keep them apart by keeping them at different corners of the property as they usually don't range too far from their houses. If they still get together, fence in a run for your turkeys. Avoid ranging them on poorly drained land as the wetness may contribute to disease problems. Lock up your flock at night so they are not threatened by bad weather or predators. In warm weather they can be outdoors at 8 weeks of age; in more severe weather wait until 10–12 weeks or until the weather has settled.

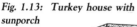

Fig. 1.13: Turkey house with sunporch

FEED

An average turkey will consume about 75 pounds of feed from birth to butchering, but you can be consoled by the fact that it has an excellent (4:1) food conversion ratio.

Poults

From birth to the age of 8 weeks young turkeys need a 28 percent protein mash fed free choice. Mash must be fed up to 4 weeks; thereafter, a pelleted feed may be provided. Supply grit early and at all other times when a whole gain is fed and the birds are confined and don't have access to natural grit. Water should always be available. Young birds may need to be taught to eat. Tempting food such as coarser-milled chicken scratch or shiny marbles will usually attract them to the food and get them started.

Growing

From 8 weeks on until slaughter the turkeys require a 20 to 22 percent protein growing ration. This comes in mash or pellet form and should be fed free choice along with some grain (corn is best, but anything that is palatable will be fine), also on a free-choice basis. Up to 16 weeks whole grains should be cracked, but can be fed whole thereafter. Greens can make up a large part of a turkey's diet. If you don't range your birds, supply them with *fresh* greens once or twice a day. Free-ranging birds receiving fresh air, sunshine, and plenty of greens will require a less complete feed (no vitamin supplements needed) than birds raised in confinement.

If a lot of greens are fed to your turkeys, it is advisable to finish birds before slaughter. For 2 weeks prior to slaughter they need only a 16 percent protein ration. Corn is the best for finishing and should be supplied free choice.

Supplementing Commercial Feed

Greens, available to free-ranging birds or supplied to those in confinement, can supply up to 25 percent of nutritional needs of your turkeys. Good greens include Swiss chard, alfalfa, tender grasses, grain sprouts, rape, lettuce and cabbage.

For the size flock most people will have, it would hardly be worth the trouble to assemble your own mash as you might with chickens. Even if you did make your own chicken mash, it wouldn't be interchangeable because of the lower protein content. You can, however, make good savings by growing some of the grains fed during the growing period.

Surplus milk can be mixed with the mash, but make sure it is cleaned up quickly and does not sour. Although I have never tried it, you can try feeding turkeys high-quality food scraps (equal to Grade I foods; see chapter on *Pigs*) much as you would chickens. Because they demand a higher protein ration than chickens or pigs, the scraps should be rich in high-protein food such as meats and cheeses. Be sure that the feed is fresh and the birds don't sling it around the cages. Spoiled feed will invite predators and disease. It is a good idea to feed such scraps only to birds on range for reasons of cleanliness.

MANAGEMENT

Routines

Since colder weather may be problem in raising turkeys, a spring to fall program is recommended. Buy your poults as soon as the weather moderates in the spring. Buy a half-dozen and figure on keeping some for yourself and selling the rest. You should have no trouble selling them and might even turn a bit of a profit. Hens bring a slightly higher price than toms.

Handling

Much the same as with chickens. If you plan to "herd" them into a corner and grab one, be careful as this may cause them to panic, stampede, and injure themselves.

Predators

Predators are much the same as for chickens. A well-constructed, predator-free cage that you can keep the flock in at night will be your best insurance.

BREEDING

Breeding turkeys is beyond the scope of most homesteaders (and this book). If you do want to try it, information is available in a good poultry book, such as Leonard Mercia's *Rasing Poultry the Modern Way*.

HEALTH

Strict sanitation and keeping your birds dry and free from chills will go a long way to keeping your flock disease-free. Do not let the turkeys mix with chickens and thoroughly disinfect any equipment that has been used by chickens in the 3 months before using it for your turkeys.

Medicated turkey feeds (which includes most, if not all, commercial feeds), if you're not against using them, should be quite helpful in preventing the two most common diseases, blackhead and coccidiosis.

The Poultry Disease Chart (see Appendix B) should be helpful in spotting and controlling most common diseases.

BUTCHERING

Butchering is the same as for chickens except that the oil sac on the back near the tail should be removed as it can cause a peculiar flavor.

2

RABBITS

One ten-pound doe can produce in her litters up to 120 pounds of meat per year—a production of over 1,000 percent of her body weight! To give you an idea of what this means, consider a good sow weaning two litters of ten 25-pound pigs per year. This would be a production of about 100 percent of her body weight. In order to equal a doe's output she would have to wean 200 such piglets or 5,000 pounds of piglets!

Rabbits make an excellent backyard venture. They do not require a great deal of care, and they are efficient converters of feed to meat. The litters are butchered before weaning, so there is no need to go the additional time and expense of building more hutches. If you have labored over the job of plucking chickens (and inhaling stray feathers), butchering rabbits will be a welcome change. A litter of 8–10 rabbits can be dressed in an hour or two. The meat is similar, though I think superior, to chicken. It is all white and has more delicate flavor. It exceeds the protein content of beef, pork, lamb, and chicken, but has a lower percentage of fat, cholesterol, and, ounce for ounce, fewer calories. The finished product is ready even sooner than veal, and because of the small bones, there is less waste (only 20 percent) in butchering than any other animal we will cover. As the world food picture grows gloomier, the rabbit may emerge as the animal of the future.

BREEDS

There are more than 100 breeds and varieties of rabbits in this country, but we need only be familiar with a small number. Rabbits can range in mature weight from 2 to 3 pounds for a Polish to up to 20 pounds for a Flemish Giant. There are four weight classes of rabbits, as shown in Table 2.1.

Table 2.1: Weight Classes of Rabbits and Common Breeds Within Each

Giants	Medium Weight	Small	Dwarf
Flemish	New Zealands	Tans	Polish
Checker Giant	Satin	Dutch	Netherland
Chinchilla Giant	Champagne D'Argent	English Spot	Himalayan
	Californians	Havana	
	Chinchilla		

While one might think the giant breeds would be the logical choice for meat breeding, they are too heavy-boned and thick-skinned and too poor in feed conversion to be economical for meat use. The medium-weight breeds are light-boned and thin-skinned so less food goes into the production of skin and bones and more into meat (Fig. 2.1). Most breeds within this class produce rabbit-fryers of 3 to 4 pounds in 8 weeks. The small and dwarf rabbits are most often used for show and laboratory purposes.

Fig. 2.1: Examples of rabbit breeds

New Zealand White

Californian

The hides of the medium weight breeds, although not as high quality as the fur breeds, are nonetheless suitable for home use. Your choice of medium-weight breeds will probably be limited because in most parts of the country rabbits have yet to be accepted on a large scale as meat producers. The most popular and available rabbits are the New Zealand Whites. They are hardy and fast gainers and therefore the most desirable.

PURCHASE

If, after you decide on a breed, you can't locate a source in your locality, contact your county extension agent or write the American Rabbit Breeders Association for sources of your choice in your area. You might also check agricultural market bulletins or look for rabbit clubs. If you still cannot locate a breeder, check in the back of the rabbit publications listed in the Appendix for mail-order services. This is more expensive and your chances of loss are higher, but it will give you a wider selection of breeds.

Age

Since rabbits don't reach maturity until 5 or 6 months of age, if you buy weaned rabbits (8 weeks old or so), you'll have to wait before you can begin breeding. On the plus side, they are cheaper than adult rabbits, and by buying young stock, you will ease into the rabbit business, giving yourself extra time to iron out any problems before you have to worry about breeding. Buy young stock as early in the spring as the weather will permit so they will be old enough to breed in the summer. Cold weather breeding can be a problem, especially for first litters.

If you are ready to go into breeding, try to buy some 5- or 6-month-old stock, but be prepared to pay a slightly higher price. The age of rabbits is difficult to determine accurately so you'll have to rely on the word of the seller (or try to get a look at a particular rabbit's hutch card if there is one). You won't run the risk of an overage animal if you buy from a friend or reliable dealer. Older rabbits should be avoided, because if an older doe has not been mated for some time, she will often be sterile. Your older bargain-basement doe may never produce a litter. Old rabbits can often be spotted by their long and heavy toenails.

Start with one buck and two or more does. If you plan to have more than 10 does, you will need 2 bucks, as one buck is able to service up to 10 does.

Sexing

Determining the sex of a young rabbit is difficult for the novice. To sex a rabbit, place it on its back and use the index finger of one hand to hold the tail out of the way. Using the thumb and index finger of the other hand, firmly but gently press down and manipulate the sex organ so that you expose the reddish mucous membrane (Fig. 2.2). In bucks the membrane will protrude to form a circle, while in does it will form a slit with a small depression toward the rear. An easier, although less reliable method of sexing is to hold the rabbit by the scruff of the neck and feel along the belly. If the area is smooth, it is a doe; if it's bumpy, it's a buck. In older rabbits, the testicles will be prominent in males, so sexing presents no problem. With enough practice you will be able to sex a rabbit quite easily. In the manipulative technique, do not use one rabbit too long as irritation may result.

Fig. 2.2: Sexing rabbits

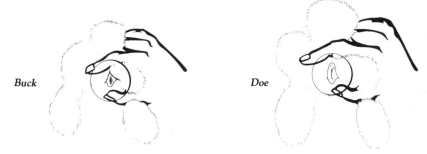

Buck

Doe

In selecting breeding rabbits for meat, look for ones that are chunky and blocky as opposed to the long, rangy ones. They should have large feet without any sores on them. Similarly, the ears should be free from sores and the mouth should be well formed and not have teeth that are broken or "buck teeth" (malocclusion). The rabbit should appear alert and have bright eyes and a smooth, glossy coat. Inquire about the rabbit's parentage as their characteristics were passed on to the rabbit and will in turn be passed on to its offspring. All large rabbitries have extensive hutch records that you can look at to help you in your choice. (This is another good reason to buy from an established rabbitry.)

Look for stock that comes from large litters whose mother consistently weaned 7–10 young. Few deaths prior to weaning would also indicate a good milk supply as well as disease resistance. Studying a hutch card will also give speed of growth as well as the number of litters per year and other pertinent data. Remember, as with other stock, it is better to settle for the worst from a good litter than the best from a poor one.

HOUSING

Rabbit hutches should be placed where the rabbits will be protected from rain and snow, windy conditions, and neighborhood dogs and children.

Rabbits do not require a great deal of sunlight, so they can be raised in a barn or garage. If they are kept outside, they should not be exposed to too much sunlight, particularly in warm weather. Sunlight helps keep the sanitation at a satisfactory level, but whether it helps the rabbits hasn't been determined.

The flooring under the hutches should be concrete or dirt. A wood floor will absorb urine, thus creating an unsanitary condition. A concrete floor requires frequent cleaning. Wood chips, sawdust, or peat moss under the hutches will absorb moisture and minimize odors.

After years of trial and error with various materials, commercial growers have settled on all-wire hutches. They are inexpensive to build or buy, lightweight, will last for years, require little cleaning, and provide a healthy environment for the rabbits. Cages built with wood and wire absorb urine and are gnawed by the rabbits.

Farm supply stores stock these wire cages. But if you plan to have a dozen or more breeding rabbits (and you need a cage for each one), it is far cheaper to build your own hutches. Even the person with 10 thumbs can build very satisfactory ones.

The following instructions are for building a hutch that is 36″ × 30″ × 18″ high. This is large enough for a medium-sized rabbit and her young.

The Wire Hutch

Tools:
>Tape measure or yardstick
>Hammer
>Pliers
>Wire cutters
>J-clip pliers
>One piece of 2 × 4 or 2 × 6 board, about 2′ long

Materials

> 1 piece 14 gauge 1″ × 2″ welded wire fencing, 18″ wide, 11′ long
>   (four sides of hutch)
> 1 piece 14 gauge 1″ × 2″ fencing, 30″ wide, 36″ long (top)
> 1 piece 14 gauge 1″ × 2″ fencing, 12″ × 14″ (door)
> 1 piece 14 gauge 1″ × ¹/₂″ fencing, 36″ wide × 42″ long (bottom)
> 80 J-clips
> Door latch

Forming the sides: First, take the natural curl out of the 11′ piece and the top. Place each piece on the floor so that the center, not the ends, tends to rise. Place the 2 × 4 or 2 × 6 on one end, stand on it, then gently pull the other end of the fencing toward you enough to remove the curl.

Turn the side piece over. Now, as you bend it up to make the corners, the bend will not tend to strain the weld but will push against it.

Measure 36″ from one end, place the piece of wood at that point, lift the 36″ section up against the 2 × 4, then tap each wire with the hammer to form a 90-degree angle. Continue to create, in turn, the 30″, 36″, and 30″ remaining sides. Using the J-clip pliers and J-clips, fasten the two ends to form the box, placing J-clips every 4″.

Forming the Bottom: Do not flatten this 36″ × 42″ bottom piece unless it is very tightly rolled. Its natural curl will help to prevent the floor from sagging.

This bottom piece is cut large enough to provide both a bottom and four 3″ sides of this tighter wiring so tiny rabbits won't fall through and out of the hutch.

Cut a 3″ square from each corner of this piece. Check to see that the ¹/₂″ wires are on top where the rabbits will be standing; then, using the hammer and piece of 2 × 4, bend the 3″ sides up. Fasten this bottom to the sides, using J-clips at the corners and every 4″.

Putting on the top: Put the top in place and fasten it with J-clips every 4″.

Cutting the door: If you're right-handed, you'll find it more convenient to have the door 4″ in from the right side of the 36″ front of the hutch. If you're left-handed, place it on the left side. In either case, cut it 4″ from the bottom and make the doorway 12″ wide and 11″ high.

Don't cut the wires flush. Leave a ¹/₂″ stub, then use the pliers to bend the stubs around the adjoining wire. That way, there's no sharp projection to scratch the rabbit or your arm.

Installing the door: A latch will be needed. Small latches made especially for wire doors can be purchased, or you can improvise with a dog leash clasp, binder ring, or even an oversized paperclip. If you purchase a latch, install it now.

Install the door, centered over the doorway space, and hanging so that it will swing inside and up. This is the better way to hang it for several reasons, among them, that when you forget to hatch it, the resident rabbit may not discover your error. Use five J-clips, leaving them quite loose so the door will swing.

For convenience, you may want to fashion a small hook from coat-hanger wire and hang it from the top of the cage to hold the door open.

Placement of the hutch: This hutch can be used in several ways. It can be placed on a dirt floor, raised on four wooden legs (2 × 2s are fine), suspended by light chains from beams, or, as many commercial growers do, mounted two-deep on an angle iron frame built to hold a dozen or more hutches.

If one hutch is mounted over another, the lower cage should be protected from the one above by a sloping piece of light galvanized sheet metal.

This hutch is particularly well designed for the rabbit raiser who is going to build a large number of hutches and so can take advantage of buying fencing at the cheapest rate, in complete rolls of the required widths. In estimating your needs for several hutches, remember that 2″ of fencing are lost each time you cut the 1″ × 2″ of fencing.

Fig. 2.3: Building the rabbit hutch

This hutch is built with three pieces of fencing, the top, the four sides, and the bottom

To form corners, wire pieces are held against board, then individual
wires are tapped with hammer

J-clips hold wire sections together

To form 3″ sides extending up from bottom, first cut 3″ squares
from each corner of the bottom piece

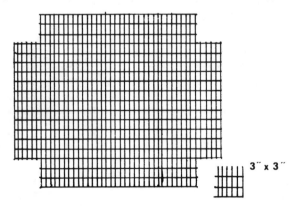

3″ x 3″

For safety's sake, hang door so that it swings inside and up

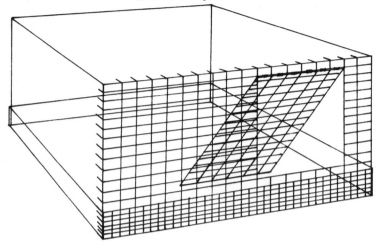

EQUIPMENT

The hutch we have just discussed will provide clean and comfortable conditions for the doe and her young and is ideal for the person who wants to minimize the work involved in feeding and watering the rabbits. Careful planning and the right equipment can cut the time for this daily chore to only a few minutes.

Feeders

You can use tin cans or crockery feeders for pellet feeding. With them, of course, you have to open the hutch each time you feed the rabbits, and much feed will be wasted as a result of tipped-over cans and contamination by droppings.

Fig. 2.4: A self-feeder that can be filled from outside and that has a lip inside that keeps rabbit from scratching pellets out and thus wasting them

Ideally, you will buy self-feeders. They are not expensive and will save their cost in feed not wasted. The feeder is mounted on the front of the hutch so that the trough extends through a hole in the fencing into the hutch while the hopper is on the outside, where it is easily filled.

Similarly, a hay feeder can be built so that hay can be inserted from the outside and is within the reach of the rabbit. Use pieces of 1″ × 2″ fencing to build a box with an 8″ × 8″ bottom, 6″ × 8″ sides, and a 6″ × 8″ back. Using J-clips, build this box, then attach it to the top, with the two sides attached to the front of the hutch. An ideal location is centered above a self-feeder. Cut the wire in the front to provide a way of putting hay into the feeder. Bend stubs around the adjoining wire to avoid scratches.

This hay rack has an advantage over hay racks mounted on the side of the hutch. Any hay dropping out will fall into the hutch and be eaten by the rabbit, rather than falling on the floor outside and being wasted.

Fig. 2.5: Hay rack that lets rabbit pull hay down into hutch

Waterers

Keeping a supply of water within reach of your rabbits is a must. Many who keep rabbits use cans or crocks as containers. These have disadvantages. The rabbits will foul them or spill out the water.

One alternative is to purchase a tube bottle waterer with a drinker tube. This is held to the outside of the cage with hooks and wire, and the tube extends into the hutch.

You can make a very satisfactory waterer using the largest size soft drink plastic bottle. Buy a drinker valve (used for automatic watering systems) and mount the valve near the bottom of the bottle by cutting a hole for it, then setting it firmly in place with epoxy cement. Hold it on the outside of the hutch with wires placed so they can be slipped off when the bottle must be refilled. Position the bottle so the valve extends into the hutch.

Both semiautomatic and automatic systems can be installed, using the drinker valves and black plastic pipe. The semiautomatic system requires a tank placed high so the system is gravity-fed.

The automatic system is linked to a piped water supply. It requires some method of reducing the water pressure before the water enters the plastic pipe. One method is to pipe water from a tank, with the intake to the tank controlled by a float valve such as is found in a toilet tank.

Both of these systems will freeze in cold water unless protected. One method is to drain the entire system, a tiresome job if it has to be repeated many nights. The other, more satisfactory method, is to insert electrical heating cable inside the pipes.

Fig. 2.6: The tube bottle waterer, left, and bottle with drinker valve

Nest Boxes

Next boxes are essential when the young are born and until they are strong enough to venture into the hutch.

The first rule of building is to make sure a nest box is small enough to fit through the 12″ × 11″ hutch door since it must be put in for each litter and taken out when the little rabbits no longer need it. A wood box 11″ × 18″ × 8″ high can be built. To avoid having the rabbits gnaw the wood, all edges must be covered with metal. Moisture holes should be drilled in the bottom.

Better than the wood box is a wire box of the same dimensions, but with the 8″ front cut down to only 4″ high. Make this of 1″ × ½″ galvanized wire fencing.

Cut one piece 18″ × 27″ for the bottom and two sides, one piece 11″ × 8″ for the back, and a final piece 11″ × 4″ for the front. Bend the sides and bottom using the method explained in the construction of the hutch. Attach the pieces with J-clips at all corners and 4″ apart.

The four top edges of the box should be covered with flanges to avoid hurting the little rabbits. Make these of 3″ wide pieces of galvanized sheet metal of the right length. Bend the strips around a ¹/₂″ wooden dowel, then hammer the metal to form a small circle atop two ends. Fit these flanges over the top edges of the back, sides, and front.

Complete this box by cutting a piece of heavy cardboard the size of the bottom. Poke drainage holes in it. Place it in the bottom of the box, then cut a piece of the fencing the same size, and place this over the cardboard. This will keep the mother rabbit from chewing down through the cardboard.

Place several inches of shavings plus lots of straw in the box. These are building materials for the doe, who will add to the pile, taking fur from her own body.

In cold weather, fashion an insulation lining for the four sides of the box, using pieces of cardboard. This cardboard and all the other nonmetal material in the box should be replaced after every litter.

Fig. 2.7: A nest box large enough for a doe and her young—and small enough to fit through hutch door

Shaping the metal strip around the dowel produces a keyhole shape for the flange

Miscellaneous Equipment

A salt spool, available at feed stores, should be furnished in each cage. Also, a small block of wood thrown into the cage for the rabbit to chew on may prevent chewing on exposed wood. It also helps to keep their teeth in proper condition.

Last but not least, the hutch card. Since it is strongly recommended that you keep records for *all* your livestock, this should not come as a surprise. The card, located either on the hutch or in a record book in your home, should list such information as name of animal, date bred and by whom, kindling date, number born, number weaned, weight, and other data. On a buck's card include all pertinent information on breeding and his offspring. These records will enable you to select good breeding stock and cull poor producers. (Such cards are often supplied by feed stores.)

Tattooing is the best method of identifying rabbits. Ear tags and clips are not suitable as they invariably tear out. Tattoos and tattoo pliers are available from most large livestock supply houses.

FEED

Commercial rabbit pellets are higher in protein (18%–20%) than most other feeds and are, accordingly, more expensive. However, the rabbits' good feed-to-meat conversion ability (3.5:1) makes it possible to raise meat economically using only commercial feed.

There is as much danger in overfeeding stock as there is in underfeeding. It is rather difficult to set any hard and fast rules for feeding rabbits, since their requirements vary with breed, size, age and weather conditions. Generally speaking, a mature rabbit that is healthy and not pregnant will require 3.8 percent of its body weight in feed daily. Therefore a 10-pound doe will eat about 6 ounces of feed per day (10 × .038, or .38 pound). Because of all the variables mentioned above, this calculation should serve as a rough guideline so that beginners don't give a rabbit 10 pounds of food a day or expect it to subsist on 3 pellets. As you learn more about rabbits and your stock, you will be able to tell if they're getting the correct amount. If possible, have an experienced breeder show you a properly maintained animal. Get the feel of it, especially around the ribs and backbone. If a rabbit person is not available for advice, pick out one of your own rabbits that is producing well and study it. If the flesh over the backbone and ribs is spare, increase the feed. If there is too much meat, reduce the feed and continue to experiment until you reach the optimal level. In time, with daily handling of your herd, you will be able to maintain your rabbits well and to spot quickly and correct any problems.

There are two types of pelleted rabbit feed: all-grain pellets designed to be fed with supplemental hay, and complete pellets containing all nutrients necessary for a completely balanced diet.

Junior does and bucks, mature dry does, and herd bucks not in service but in good physical condition can be maintained on hay alone if it is a fine-stemmed, leafy, and legume variety. If coarse grass hay is the only type available, it can be fed if each 8-pound animal is supplemented with 2 ounces of an all-grain pellet several times a week. Adjust the amount of grain for rabbits of varying weights. For example, a 4-pound rabbit would require 1 ounce of pellets while a 12-pound animal would require 3 ounces.

Alfalfa pellets may be fed to developing junior does and bucks as their only feed from the time they are weaned until they are ready for breeding. Buy pellets containing 1 percent salt and 99 percent of no. 2 leafy or better-grade alfalfa meal (15% protein).

Feeding Does/Creep Feeding

In feeding pelleted feed to pregnant does and does with litters, allow them all they will clean up between feedings. Creep feeding—placing feed in a container accessible only to the young rabbits—prior to weaning is definitely not necessary but is gaining some favor among rabbit raisers as a method of quicker and more substantial gains (see Chapter 3, Fig. 3.3 for lambs). The doe's milk is the perfect food for young rabbits and is all that is needed until slaughter. Its high protein content (15 percent) makes it one of the richest milks of all livestock. If you want to experiment with creep feeding, supply pellets free choice from a creep feeder. At least one company, Carnation, makes a creep feed especially for young rabbits that is higher in protein (22 percent) than most rabbit feeds. Again, as with all animals, experiment, comparing weights of those litters that are creep-fed and those that are not, and decide whether the added cost of feed is worth it.

You can feed your rabbits as many times a day as you want, but one feeding is adequate as long as it's enough. *Be consistent.* Rabbits eat the most in the evening and night, so if only one feeding is supplied, give it in the evening.

Contrary to popular belief (reinforced by cartoons of Bugs Bunny subsisting only on carrots), rabbits cannot live solely on carrots and greens. If you feed greens (see below), allow your rabbits only what they will clean up in 10 to 15 minutes in the evening.

Water/Salt

As mentioned earlier, a mother and her litter must have water available at all times, for they may consume up to 1 gallon of water in a day. Fresh water should *always* be available. Salt, in the form of a salt spool, should also be provided free choice.

Supplementing Commercial Feed

In conventional feeding, providing only commercial feed, you have a complete food and won't have to worry about other nutritional needs. In supplementary feeding you must have some knowledge of rabbits' nutritional requirements to be certain that they are met. In simplest terms, does (pregnant or with litters) need a 20 percent protein ration. A pelleted feed supplies this, or it can be provided in the formula provided in Table 2.2.

Dry does and bucks need about an 8 percent protein ration. They can, of course, be fed commercial feed but they are getting much more protein than they need (20%). It won't hurt them—only your pocketbook. You can feed other things, as we shall see, but the general formula is given in Table 2.3.

Commodities that make up each of the above groups are given below. Some of

Table 2.2: Feed Requirements for Does

	% Ration (by Weight)
Protein	20
Grain	39.5
Roughage	40
Salt	0.5
	100.

Table 2.3: Feed Requirements for Dry Does and Bucks

	% Ration
Protein	8
Grain	31.5
Roughage	60
Salt	0.5
	100.

these you might be able to grow, or come by cheaply; still others you will have to purchase commercially. In any event, making use of the above formulas and experimenting with combinations will enable you to decide on a food mix that is the best for your stock and is the most economical for you. Sources:

Protein: In order of palatability and nutritional value for rabbits are peanut meal, soybean meal, linseed oil meal, hempseed meal, and cottonseed meal. Whole soybeans, while high in protein, are often less palatable for rabbits. Pellet meals or pea-sized cakes of the above meals are best.

Grains: Cereal grains in order of desirability are oats, wheat and grain sorghums, barley, corn, and milo. Barley and oats need not be rolled. Bread or other wheat products can be used.

Roughage: Any good legume hay (second cutting is best). It is eaten most efficiently if cut into 3- or 4-inch pieces. Carbonaceous hays such as Sudan grass and timothy can be fed, but the protein content of the feed should be increased a bit.

Salt: Salt can be mixed in with the feed at the indicated rates. Better, supply a mineralized salt spool for all your rabbits.

The Walters family of Campbellsport, Wisconsin, has developed the feed program given in Table 2.4 and has reported good success with it. It may not perfectly suit your rabbits, but it can be used as a starting point in formulating a more self-sufficient ration for your rabbits.

Sally Cook of Rochester, New Hampshire, reports success with raising rabbits on a program that includes millet, squash, and apples. She grows a plot of millet that she feeds to her herd unthreshed and unwinnowed, allowing them to eat the hay along with the grain, which the rabbits husk themselves. She collects culled or damaged squash from her garden and from a nearby squash farm for winter feeding. She allows her rabbits as much of it as they will clean up in a day. They will eat the pulp but enjoy digging for the high-protein seeds inside. Apples that are picked in abandoned orchards and are left over from her cider-making are fed during the winter. While she does not gain top efficiency and has only five or six litters per year, she maintains that her herd is healthy and provides her with good meat at a minimum cost.

Table 2.4: Walters Feed Program

Summer

Dry Does and Bucks:

Morning and evening—Greens and weeds from garden (see *Table 2.6* on Desirable and Undesirable Greens) No more than they will clean up in 10–15 minutes.

Evening—Grain supplement (4–6 oz. of raw grain consisting of 3–5 oz. oats and 1–2 oz. soybeans)

Pregnant Does and Does with Litters:

Morning and Evening—Weeds and greens (see above)

Evening—2–2$^1/_2$ cups grain mixture consisting of 30% corn (omit in warmer months), 60% oats, and 10% soybeans.

Fall and Winter

Hay and a grain mixture of 45% corn, 45% oats, 10% soybeans. Feed free choice to active does and enough to maintain dry does and bucks.

Other rabbit feed recipes that can be made up by the industrious homesteader, as supplied by the U.S. Department of Agriculture, are given in Table 2.5.

To make a feed mix, assemble all the ingredients in the correct proportions and feed it to your rabbits. If they get picky and eat only "favorite" parts, have the ration ground. Whenever you feed mash to rabbits, be sure to wet it down a bit as dust will interfere with a rabbit's breathing.

Since rabbits eat very little compared to pigs, any food surpluses you use will result in substantial savings. If you can grow corn and have access to stale bread, you have two good groups accounted for (using the food program in Table. 2.4) and need buy only a protein supplement to complete your ration. Often farmers will, for the asking, allow you to glean their fields after harvest in the fall. People I know collect left-behind ears and for a few days' work have all the corn they need for a winter.

To get corn off the cob, you will need a corn sheller that is available from feed stores or livestock supply catalogs. Oats need not be rolled as rabbits will shell them themselves, but rolling will prevent waste. For rabbits that will eat it, the chaff can make good roughage. There may be more waste when feeding whole grains, so it's a good idea to place a board under the feed dish to collect dropped food.

Table 2.5: Rabbit Feed Recipes

For Pregnant Does with Litters		
(1)	Whole oats or wheat	15%
	Whole barley, milo or other grain sorghum	15
	Soybean or peanut meal pellets or pea-sized cakes (38–43% protein)	20
	Alfalfa, clover, or pea hay	49.5
	Salt	0.5
		100.
(2)	Whole barley or oats	35%
	Soybean or peanut meal pellets or cakes	15
	Alfalfa or clover hay	49.5
	Salt	0.5
		100.
(3)	Whole oats	45%
	Linseed meal pellets or cakes (38–43% protein)	25
	Carbonaceous hay	29.5
	Salt	0.5
		100.
For Dry Does, Bucks, and Young Rabbits		
(1)	Whole oats or wheat	15%
	Barley, milo, or other grain sorghum	15
	Legume hay	69.5
	Salt	0.5
		100.
(2)	Whole barley or oats	35%
	Legume hay	64.5
	Salt	.5
		100.
(3)	Whole oats	45%
	Soybean, peanut or linseed pellets or pea-sized cakes	15
	Carbonaceous hay	39.5
	Salt	.5
		100.

Vitamin supplements to supplementary feed are generally unnecessary because the vitamins A, D, E, and B should occur naturally in all feeding routines.

Vitamin A is present in some root crops (carrots, Jerusalem artichokes) and good hays. Vitamin D is present in hay that is sun-dried, and vitamin E is present in dry hay, grains, and protein supplements. Vitamin B is synthesized in the stomach of the rabbit in a type of pseudo-rumination, and the vitamin is obtained by eating a small amount of mucuous-coated feces that is passed at night.

As stated earlier, when fed with care, greens can supplement feed. Young rabbits should never be fed greens as they may fill up on them and ignore more nutritious food. In a limited way greens are quite helpful, but overingestion must be watched because of their high water content. Table 2.6 lists good and bad greens.

Table 2.6: Desirable and Undesirable Green Feeds and Root Crops for Rabbits

Desirable	
Rapidly growing cereal grains	Jerusalem artichokes
Rape	Gren hay crops
Trimmings from leafy garden vegetables	Kale
Dandelions	Lawn grasses
Sweet potatoes and vines	Tender twigs
Plantains	Filaree, afilaria, or stork's bill
Mangels	Malva or mallow, cheeseweed
Swedes	Carrots and tops
Kohlrabi	Turnips
Culled potatoes and peelings	Sugar beets

Undesirable	
Burdocks	Miner's lettuce
Castor beans	Nightshade
Fireweed	Oleander
Goldenrod	Poppies
Horehound	Sweet clover
Lupine	Tarweed
Milkweed	Rhubarb

Source: From *Domestic Rabbit Production* by George S. Templeton. Danville, Il: Interstate Printers & Publishers, Inc., 1968, p. 59.

MANAGEMENT

Routines

While not too many Americans are familiar with rabbit meat, once they have tasted it, they usually become converts. Invite some friends over for a rabbit dinner, and I'll bet they will become a ready market for your surplus meat rabbits (or for some weanlings as breeding stock) and get started themselves. By selling your surplus meat, you may be able to pay for your own meat and even turn a little profit.

As an added plus, rabbit manure is one of the most valuable manures. You can use it for your own flowers or vegetables or find a market for it with earthworm farmers—or maybe begin growing your own earthworms. The pelts, while not valuable commercially, can be put to many uses around the house for gloves, hats, and blankets. *Do not* be tempted to sell little bunnies as Easter gifts for children. While it may be seen as a source of added cash, more often than not it also ends up in death or serious mauling for the rabbits. Rabbits as Easter gifts probably are the most ill-treated of all pets.

Chewing: To prevent chewing and the consequent destruction of the wood portions of your hutches, cover exposed wood with tin or wire. Blocks of wood and twigs will help to satisfy the rabbits' urge to chew and condition their teeth.

Molting: Chickens lose their feathers; rabbits, their fur. This shedding and restoring of the coat normally takes place once a year in mature rabbits and in the first several months of age in young rabbits. It may be caused, unnaturally, by disease, sudden high temperatures, rapid changes in diet, or other traumas. Normally, molting should be a minimal concern except that extra feed will probably be needed, and conception in does and production of sperm in bucks may be adversely affected.

Handling

Carrying rabbits around by their ears will result in injured animals. It should be left to magicians who can presumably make their injured animals disappear. The correct way to pick up and carry a rabbit is by grabbing the skin of one shoulder in one hand and placing the other hand under the rump for support. The feet should always be pointed away from the body to prevent scratching.

Predators

Cats, dogs, opossums, weasels, snakes, and rats share with humans a love of rabbit meat. Housing your rabbits in a barn or other closed building will eliminate the most common predator, dogs. Well-constructed cages using wire should keep out most predators. Safeguard your young rabbits by having a thick bottom on nest boxes to prevent rats and other animals from gnawing through and getting at them. If rats or snakes climb up the legs of your pen and into the cages, you should put circular sheets of metal around each leg to block their path.

BREEDING

The age of maturity varies with the weight class of the rabbit. Medium-weight rabbits usually reach sexual maturity by 6 months but sometimes as late as 7. The dwarf breeds mature at 5 months, and in the giant breeds maturity is reached from the tenth to the twelfth month.

The doe's egg cells are contained within small follicles. In 9 to 10 hours after sexual stimulation by the buck, the follicle will break. The sperm that was deposited at the time of mating will fertilize egg cells within 2 to 4 hours after the follicle breaks.

The doe should be taken to the buck's cage for mating. This will tend to prevent fighting, as a doe is apt to attack a buck when he is taken to her cage. He will usually mount her without incident and, when the service is completed, he will often grunt and roll over. There is some controversy as to whether conception rates are improved by returning the doe for remating in a few hours. Some people say returning her will not improve conception but will only serve to deplete the buck. Others claim that since the sperm must be retained in the body until the egg follicle ruptures, the sperm may pass out of the body if the doe urinates. They think remating in a few hours (some also remate twice the next day) will ensure pregnancy. Experiment.

In some cases the doe will not accept the buck. If she ignores him or growls, do not leave them alone. They might fight and injure each other. Watch them for a while. If the doe evades him but her tail is twitching, she probably will accept him in time. If, however, the doe does not accept him, you might try moving her cage next to his for a few days so they can get used to each other. Often she will accept him when he is reintroduced. Another method which often succeeds is to switch cages. After a few days, take the doe back to her cage, leaving the buck inside.

If you want results *immediately*, you can try force mating. Grasp the doe by the ears and the skin of the shoulder with one hand and place the other underneath her with two fingers on either side of the vulva, pushing it out a bit. As the buck mounts her, raise her hindquarters a bit to allow normal coupling. This may not always ensure conception as there are four days in the doe's 16-day cycle when no eggs are available for conception. During these four days she also lacks interest in the buck, and this may be why she rejected him.

The gestation period is an average of 31 days—as short as 29 and as long as 35. Three to five days before kindling, or giving birth, place the nest box and bedding (straw, hay, but not sawdust as it can asphyxiate the young) in the hutch. Kindling is imminent when she begins plucking her fur to line the nest box.

The day after kindling, offer the mother rabbit a tempting morsel of food, then quietly remove the nest box and inspect the young. Remove any dead or deformed babies and any runts in a large litter. At this time you may even out litters by taking some babies from a large litter and letting a doe with a small litter nurse them. This can be done with litters which have kindled up to 3 days apart and is a good reason why you should breed your does to kindle about the same time. While litters may range in size from 1 to 20, 8 to 10 is the ideal number for a doe to handle.

At about 10 days of age their eyes will open and soon the babies will be in and out of the nest box. They will begin eating some of the doe's food and if you want, they can be creep-fed.

At 8–10 weeks of age they should be weaned and slaughtered or some retained for breeding purposes. The nest box can be removed at 4 weeks or may be retained longer in colder weather.

Problems

Your doe may kindle her young on the floor outside of the nest box, or refuse to nurse her young, or kill them. Usually the cause is simple, such as moving her to a new cage too soon before kindling, excessive noise, or the presence of predators. Does are often quite sensitive close to kindling time and so it is wise to leave them alone as much as possible. If you find young on the floor of the cage, don't panic. When I was young, one doe of mine kindled 8 on the floor of the cage, but upon closer inspection I found that there were also 10 in the nest box. I put them back in the nest box and she just as promptly returned them to the floor of the cage. Apparently, as is so often the case with livestock, it was a case of not giving her credit for doing what nature told her. She was evening off the litter herself as she apparently knew if she tried to nurse all 18 they would die or at best be so weak that they would fall victim to disease.

If does fail to pull their fur off if they kindle on the hutch floors, make a comfortable nest and pull enough fur from the doe to cover the litter. Keep some extra fur on hand for such instances.

Many times it is possible to save litters that appear completely lifeless by warming them.

Occasionally the eyes of baby rabbits may become infected and fail to open normally. If these infections are treated promptly, the young rabbits generally recover without any permanent eye injuries. If the eyes appear to be infected, bathe them with warm water until the tissues soften and the lids can be separated with slight pressure. If a pussy discharge persists, an antibiotic eye ointment should be obtained from your veterinarian.

If a doe won't nurse and you find there is no milk, mastitis may be the problem. This occurs as a result of bacterial infection or injury to the mammary glands. At the onset, the udder will be pinkish and hot and the teats may be pink or light blue. Wash and disinfect all equipment and the nest box. For treatment to be effective (100,000 units of penicillin injected in the thigh), it must be carried out at this point. Death may result if treatment is not given promptly. Let her nurse her young if she can or will. Do not resort to a foster mother or the infection may be spread to her.

In the case of orphaned litters or when a doe won't nurse, you can foist some on another doe if they were born within 3 days of hers. To prevent her from rejecting them, put some Vaseline on her nose. In cases where you have no other doe to pass orphans or "extras" from a large litter onto, you *can* raise them yourself but, please, think twice. Raising any animal from birth is a taxing job, and rabbits are among the most demanding. They must be fed with an eyedropper or doll bottle at least 4 times a day (and at night). If you're not sure you'll stick with it, drown them and dispose of them. This is preferable to stringing them along for a few days, slowly starving, and then deciding the drudgery is no longer worth it.

If you decide to raise them yourself, use cow's or goat's milk or evaporated milk mixed with water to the consistency of whole milk. Heat the milk so it is warm on the back of your hand and feed it with an eyedropper *at least* 4 times a day (8 times a day is preferable for the first 4 days). At two weeks they can be offered rolled oats and a few blades of grass. By 15 to 18 days they can be taught to drink from a saucer and eat regular rabbit feed. Moistening the feed slightly will make it more palatable.

It is possible, even after you witness mating, that the doe will not conceive. There are a variety of reasons: sterility in the doe because of age or a long period of nonproduction, sterility of the male, disease, molting, she is not in heat, or because of false pregnancy. False pregnancy occurs when a doe is mounted by another doe prior to breeding. This ruptures the egg follicles but obviously no conception will take place. False pregnancy results and conception by a buck is impossible for 17 days. This is good reason not to house does in the same cage.

During false pregnancy the doe goes through all the routine that she would go through if she were actually pregnant, including nesting.

Because of the above cases of nonconception, it is helpful to be able to tell if a doe is pregnant earlier than the thirty-first day, so you can adjust her feed and rebreed her as soon as possible. You can feel the young most easily in the doe on the fourteenth day after breeding. (After that they are too large, and indistinguishable from the internal organs by all but the most expert.) It is best to test her in the hutch so as not to unduly upset her. Restrain her with one hand, holding the ears and the scruff of the neck as in force mating; with the other hand, gently feel the abdominal area immediately in front of the hind legs. At this stage the young are about the size of marbles, and with some experience, you should be able to detect them. To practice, use a doe you know is not pregnant and compare her with one you are sure is. In time the testing will be routine. A less reliable method is to reintroduce her to the buck 2 weeks after the first breeding. If she growls or whines and otherwise rejects the buck, *chances are* she is pregnant. But this is not a sure thing—remember, even if a doe is not pregnant she may act the same way if she isn't in heat.

If she appears not to be pregnant, the doe should be bred again. Even if you're sure she isn't pregnant and she accepts the buck again (pregnant does have been known to be mounted even while pregnant), put the nest box in a few days before the original kindling date just to be safe. It won't hurt, and people have had litters from a "dry doe."

Rebreeding

In the wild, rabbits come into heat only in the fall and spring. With domestic rabbits, as mentioned earlier, there is a 4-day nonfertile period in each 16-day cycle, but otherwise conception can occur any time during the year. The most normal time to rebreed is after a kindling and 8-week nursing period. You shouldn't wait too long between weaning and mating as sterility may result. The rebreeding schedule mentioned above would enable a doe to have 4 litters a year, climate or warmth of your hutches permitting (31 days gestation + 56 days nursing = 87 days; 87 × 4 = 348). If an insulated or heated nest box still doesn't allow you to have successful winter kindling in your part of the country, you will be limited to a maximum of 3 litters a year: the first to kindle in early April (or earlier if your weather warms up before then) and the third to kindle in late September or early October.

Rebreeding can be carried out as early as the fourth week after kindling (take the doe to the buck's cage as in normal mating), allowing a possible maximum of 6 litters per year, depending on your doe. Don't try to force her or she won't last very long. If she is healthy and chunky, you can rebreed her early; if she is drained and depleted, wait at least until she's weaned the litter or until she is back in condition. With experience and the daily handling of your rabbits, you'll be able to determine what is best for your herd and your does.

In the case of a lost litter, the doe's feed should be cut to regular maintenance ration and she should be supplied with roughage to help her dry up. She can usually be remated 3 days later. When a doe has only one or two surviving young, they can be given to another doe and she can then be dried up and rebred in a day or so.

Experiments in breeding can be fun and rewarding if you begin to markedly improve your herd through selective breeding. An acquaintance of ours reported success with breeding purebred medium-weight breeds (New Zealands, Californians) to a giant breed such as Flemish or Checker to produce larger fryers maturing in the same time as pure medium-weight rabbits. Experiment!

Bucks

Your buck is half your herd and you should choose him with that in mind. Look for all the qualities you seek in your young stock. In addition, look for one with a shiny coat and bright eyes. Weight is not as critical as it is with does, but a buck that is overly fat may be lazy and not service your does. When that happens, put him on a diet and his interest should return. The scrotum should be full and large and contain two fully descended testicles. If a buck has small and withered testicles (most common in older rabbits), he may be sterile or sire small litters. Bucks of the medium-weight breeds mature in 6 to 8 months, but it is best to wait until 8 months of age before using him. The larger breeds don't mature until at least 10 months. The first few times, the buck may benefit from forced mating, especially when older does are used. As a rule of thumb, a buck can service up to 10 does. Ideally he should not be used more often than every 4 to 7 days.

HEALTH

Rabbits are among the most disease-resistant of livestock, *provided* they are well fed and live in clean surroundings. Prevention is once again the byword. Frequent cleaning and disinfecting of feed and water dishes and removal of spoiled hay, cleaning the hutch, and frequent removal of manure from underneath the pen are all good practices. A solution of lye is a good disinfectant. Blowtorches, commonly recommended for disinfecting, are not totally satisfactory because they must have direct contact to kill germs and also prolonged use may weaken the galvanizing bond on wire and invite rust.

Clean nest boxes after removing them from the hutch and drying them in the sun. Sick rabbits should be isolated for 2 weeks after symptoms disappear. New rabbits entering the flock should be similarly isolated to prevent them from bringing in any diseases. The table in the Appendix will be useful in the diagnosing and treatment of the most common rabbit afflictions.

Good ventilation is as critical in rabbit production as it is in other types of livestock grown under conditions of semiconfinement. The air in the rabbitry must be free of ammonia fumes and carbon dioxide. These noxious gases have an extremely adverse reaction on the rabbits' respiratory tract, and their continued presence can lead to severe respiratory conditions.

Rabbits need 14 hours of light daily to keep the breeders active and fertile.

The barn or hutch temperature should ideally be maintained at 50°F–75°F year round. By doing this it should be possible to raise a 4–5 pound rabbit on between 1 and 3 pounds of feed. At lower temperatures, too much feed is used to maintain body temperature while at high temperature intake is repressed.

In a small operation, achieving this range may be difficult, but effective use of fans and sprinklers should help to reduce the effects of extreme summer heat. During the winter, it is more difficult to achieve the lower end of the recommended range without affecting good ventilation.

If they show signs of excessive heat stress, animals should be moved to an area where the ventilation is excellent. They can be placed on a feed sack moistened with water, and it is also possible to place containers filled with ice in the hutches.

Sprinkling the tops of hutches with water will serve to lower the temperature 6° to 10°F on a hot day when the temperature in a nesting box can become most uncomfortable. Measures should be taken to make the babies more comfortable. Cooling baskets are wire boxes constructed of hardware cloth that can be suspended from the roof of the hutch to allow for maximum air circulation around the youngsters. At night, return the young to their nesting box, but if the outside temperature is still high, they can be returned to the cooling box.

Diseases caused by viruses

Infectious myxomatosis: This condition is a fatal viral disease of the domestic rabbit, angoras, Belgian hares, Flemish Giants, and the European wild hare. The poxvirus is transmitted by mosquitoes, biting flies, and direct contact from the California brush rabbit, which acts as the natural reservoir. The disease is restricted to the West Coast of the United States and is seen at the height of the mosquito season. It tends to flare up about once every 8 years, and then becomes somewhat quiet in the intervening years.

Animals that develop the acute form of the disease may die within 48 hours.

Animals that survive for longer periods eventually develop ears that become filled with fluid that causes them to droop.

There is no available treatment and no vaccine.

Papillomatosis: This disease occurs in two forms in the United States. One form is seen as small, warty growth under the tongue and on the floor of the mouth. The other type is seen as horny warts on the neck, shoulder, ears, and abdomen and is seen mostly in cottontails.

Diseases caused by bacteria

Pasteurellosis: Probably the most common bacterial disease of rabbits, the organism that causes it bears the imposing name of *Pasteurella multocida.* This organism can cause a wide variety of different conditions—snuffles, pneumonia, middle ear infections, eye infections, abscesses, genital infections, and septicemia.

Snuffles or nasal catarrh is an infection of the membranes of the air passages and the lungs. It can vary widely in its severity. It usually occurs when the animal's resistance is lowered for some reason. Allowed to progress untreated, it can develop into pneumonia. Animals that recover can act as carriers, accounting for the spread to other animals.

Abscesses arise particularly in bucks that are housed together and fight, giving rise in turn to abscess formation at the site of wounds which can be almost anywhere on the body. In many instances you should eliminate infected animals since abscesses may develop into an infection called septicemia, which usually results in death.

The organism may locate in the genital tract, where it causes varying degrees of inflammation. Animals of either sex that become infected with this organism have very poor futures as breeders.

Mastitis: An infection of the breasts, this disease may be caused by a number of possible organisms but in particular *Staphylococci* and *Streptococci*. Glands that are infected are hot, swollen, reddened, and painful early in the course of the disease. Later, if left untreated (or sometimes even if treated), the glands may become blue in color, giving rise to the common name "blue breasts."

Treponematosis: Caused by a spirochete called *Treponema cuniculi,* which infects the reproductive tract of both males and females, this condition is identified by first the presence of ulcers and then scabs in and on the external genitalia. Affected animals should not be used as breeders, as the organism is transmitted by mating. It is probably also spread from the mother to her young.

Penicillin is used to treat this disease. Give once a week subcutaneously. All animals in the operation must be treated whether or not they are showing visible signs. Once the scabs have healed (in about 2 weeks on average), the animals can once again be used for breeding. Penicillin may cause enough changes in the flora of the gut to initiate diarrhea. Watch the hutches for this possibility.

Hutch or urine burn: A condition that may closely resemble treponematosis is hutch or urine burn. The only specific way of differentiating the two is by laboratory examination of tissue or material from the affected areas, which are the anus and the external genitalia. This condition occurs when the hutch floors are allowed to become wet and dirty and the animal's rear quarters become irritated. Keeping the floor clean and dry will do much to prevent this condition from developing. Local application of an antibiotic ointment will assist in clearing the infection.

Enterotoxemia: An acute disease characterized by an explosive diarrhea, enterotoxemia occurs especially in rabbits 4 to 8 weeks of age. This disease is caused by an organism called *Clostridium spiroforme*, and members of this group are notorious for causing a wide range of intestinal toxemias in a wide range of animals, especially young ones. There is a relationship between diet and the development of this disease. Where diets high in fiber are fed, far less enterotoxemia is seen. This group of organisms, when conditions in the gut are just right, grow and produce a specific toxin which is highly fatal even in very small amounts.

Mucoid enteropathy: A disease of unknown cause that can be seen in rabbits of any age, mucoid enteropathy is the result of constipation. The rabbits affected show a variety of signs including gelatinous or mucus-covered feces, off feed, dull, poor hair coat, subnormal temperature, and sometimes a bloated stomach due to excess water in that organ. Treatment is not very effective, but sometimes electrolyte solutions are helpful.

Tyzzers disease: Also a disease of young rabbits characterized by a profuse diarrhea and death, this condition is caused by a bacteria called *Bacillus piliformis*, which is eaten where the level of sanitation is poor and the level of stress is high. Very likely a poor level of sanitation can give rise to a high level of stress.

Parasitic Disease

Coccidia: Rabbits can become infected with coccidia, usually at a young age. The organism can set up housekeeping in either the liver or the intestine and is usually swallowed with contaminated feed or water.

The severity of the disease in the liver form seems to be dependent on the number of organisms swallowed. The more organisms taken in, the more severe the disease. It is not generally possible to diagnose this disease until the animal dies or is sacrificed to make a diagnosis. There often are no great outward signs or symptoms that would give the owner a clue as to what might be going on except that the animals just don't seem to do well. Diagnosis is based on post-mortem findings and finding the organism in the bile ducts.

Sulfaquinoxaline is a sulfa drug that can be used either for treatment or as a preventative in the drinking water or in the feed. Check with your veterinarian for appropriate dosage levels after a diagnosis. This drug now can be administered up to within 10 days of slaughter, but keep yourself appraised as to the current regulations regarding withdrawal times. These regulations are changing rapidly at the time of this writing.

The intestinal form of the disease is acquired in exactly the same way as the liver form, but diagnosis can be accomplished by having your local veterinarian do a fecal exam. As with the liver form, the symptoms shown by the affected animals are often very vague, typical of "poor doers." Treatment is the same as for the hepatic type.

External Parasites

Ear mite: Probably the most commonly recognized external parasite of rabbits, the ear mite has a dramatic effect on the ear. The inner surface of the ears becomes coated with a thick, gummy material which arises because of the constant irritation that the mites cause, coupled with the severe beating the rabbit inflicts on itself as the result of its misery.

When attempting to clean the infected ear, it is best to heavily sedate or anesthetize the animal as the cleaning is a very painful process. A cleaning with hydrogen peroxide to remove the heavy waxy material followed by the application of a good ear mite medicine that might be used in cats or dogs is needed. It may be necessary to repeat the ear mite medicine application 2 or more time to eliminate the mites.

Mites are easily transmitted by direct contact so segregation of infected animals is essential to control the problem. Thorough cleaning and disinfection of hutches is critical to a good control program.

Non-Infectious Conditions

Moist dermatitis or wet dewlap: This condition develops when mature rabbits, especially females, get their dewlaps wet from dropping into the water bowl. When this area is allowed to remain moist continuously, it may lead to inflammation, which in turn may become infected. When this happens, it is necessary to clip the fur in the affected area and treat it topically with either an antiseptic or antibiotic dusting powder.

The use of automatic waterers will generally prevent this condition from developing.

Hair chewing and hair balls: These conditions become a problem if the amount of fur swallowed becomes excessive. Generally the rabbit is able to pass the hair through the digestive tract. If hair does become a problem, the rabbit tends to go off feed, lose condition, and die in a few weeks. Unlike some other animals, the use of mineral oil and laxatives is ineffective. Keeping the fiber level of the diet (20 percent) at an appropriately high level will keep this condition from developing.

Sore hocks: This is a problem, especially in the heavy breeds that are raised on wire floors, especially where the wire may have been improperly installed, allowing the soldered joints to protrude upward. Actually it is not the hock that is involved but rather that part of the rear foot below the hock. Urine-soaked feces allowed to accumulate on the wire adds a further complication. Rabbits with this condition tend to sit with their weight distributed toward their front quarters to take the weight off the sore, inflamed rear feet. When all four feet are involved, affected animals tend to walk on their toes.

Malocclusion: This condition occurs when the front teeth do not come in contact with each other in an acceptable way and grow too long. Usually the teeth in the lower jaw protrude outward and do not come in contact with those in the upper jaw. The teeth of all rabbits continue to grow throughout life, and the rabbits keep their teeth ground down by chewing on wood and other materials. If the problem is getting out of hand, the protruding teeth can be cut back with a pair of small wire-cutting pliers.

This condition is considered to be hereditary, so affected animals should not be used for breeders.

BUTCHERING

While some may reason that you should wait until a rabbit matures to 10 pounds before butchering, this is an unsound practice. The most efficient growth occurs during the first 8–10 weeks of life; thereafter feed conversion is not nearly so efficient. While you might get more meat by waiting, it is much more expensive.

Once you butcher a rabbit, you may never again consider eating your own chickens. The meat is delicious, better for you, and there is very little waste (20 percent). No more hours of plucking, feathers up your nose or in your shirt. Once you get the hang of it, you can butcher 10 rabbits in the time it would take to do a couple of chickens.

The first time you kill and dress a rabbit will be the most difficult. You'll need a large sharp knife and a plastic garbage bag. Drive a nail part way into a small piece of board, hang the plastic bag from it, and hang the board on a wall. You're ready to begin.

Hold the rabbit by the hind legs in one hand, as shown in Figure 2.8, then grasp the head and quickly and forcefully snap it down while twisting it back. This breaks the rabbit's neck and kills it.

Cut off the rabbit's head, then hang the rabbit by one rear foot from the nail in the board with the body over the bag. Let it bleed out completely.

Cut off the free rear foot and both front feet. Remove the tail and slit the skin (being careful not to cut the flesh) from the hock of the remaining foot down the inside of the leg to the base of the tail. Continue up to the stub of the outer leg. Now pull the skin down off the body, inside out like a glove. A knife may be used to separate the skin from the fat, which should remain on the carcass.

Slit the belly from the middle of the breastbone to the tail. Remove the entrails.

*Fig. 2.8: Positioning of rabbit
prior to breaking its neck*

Save the heart, kidneys, and liver, being sure to cut the gallbladder from the latter. Cut off the remaining foot, then rinse the carcass with cold water to remove blood and stray hairs.

As with most meat, cutting it up is easier if the meat is well chilled, although this is not absolutely necessary.

A rabbit carcass, cut up, will have 7 pieces. Cut off the front and back legs at the ball joints, then cut the back into three pieces, much like a chicken.

The hide is worth little commercially. If you want to use it, you can tan it by following the directions outlined in the Appendix.

While 8-week-old rabbits are most often butchered, those older rabbits you cull can also be used. Bucks are generally tougher than does, but they can be stewed or ground and added to pork for sausage.

3

SHEEP

While sheep may not rival many animals in intelligence, they can teach them a thing or two about feed efficiency. Like goats and cows, sheep are ruminants or multistomached animals which, through bacterial action in their stomachs, can convert cheap roughages and other inexpensive low-protein feeds into high-quality foods that are rich in protein, amino acids, and essential vitamins.

Sheep are hardy, sure-footed animals that can graze even the hilliest and rockiest pastures unsuitable for other livestock. Because they eat a wide variety of roughages, they are excellent weed killers and this, along with their rich manure, makes them an ideal animal to improve poor pasture. (They also make superior lawnmowers—much better, I think, than those outdated models that need gasoline.)

Lambs purchased in the spring will fatten quite nicely to butchering weight by the fall on grass alone. They need relatively little grazing space, and by using a tethering system and being careful, a few sheep can do well without any fencing.

Sheep will also supply you with another valuable product besides meat: from 3 to 18 pounds of wool per year that can be used for spinning, knitting, or sale to help defray the costs of your sheep and other livestock. Purchase a few ewes and a good ram and you will be supplied with lambs year after year, their cost being only a fraction of the cost of your ewe. By starting your own mini-flock, delicious lamb and useful wool will be yours for a lot less than the consumer pays.

BREEDS

There are many breeds of sheep, classified in six groups as listed in Table 3.1.

We are most interested in the medium-wool types. They are primarily raised for their meat, being rather blocky in conformation and relatively fast-maturing animals. They are not usually known for their wool production, but recently more emphasis has been placed on this in their breeding, and they are good dual-purpose sheep. If you are most interested in wool, you should look into fine-wool breeds, which are also becoming better meat producers due to selective breeding. Long-wool types are generally too slow maturing to be important here, but crossbreeding them with the fine-wool breeds has made them a fair dual-purpose animal, and for that reason they are often classed with the medium-wool types.

If you are planning to purchase grass lambs (spring-born lambs that will be raised on pasture and butchered in the fall), breed will be of little importance to you. The key considerations will be availability in your locale and personal appeal, although it is important to note that some breeds are faster maturing than others and will lend themselves more readily to a spring-to-fall rearing program. If you plan to do your own breeding, a deeper knowledge of breeds will be necessary. In any event, some familiarity with the common breeds is called for.

Table 3.1: Classes and Common Breeds of Sheep

Fine-Wool Type	Medium-Wool Type	Long-Wool Type	Crossbred-Wool Type	Carpet-Wool Type	Fur Type
American Merino	Cheviot	Cotswold	Columbia	Black-Faced Highland	Karakul
Debouillet	Dorset	Leicester	Corriedale		
Delaine Merino	Hampshire	Lincoln	Panama		
Rambouillet	Montadale	Romney	Romeldale		
	Oxford		Targhee		
	Shropshire				
	Southdown				
	Suffolk				
	Tunis				

Medium-Wool or Mutton Breeds of Sheep

Cheviot: White faces and legs covered with short white hair. Black noses, lips, and feet with short, erect ears. Cheviots are quite hardy, prolific (125 percent lamb crop*) and they make good mothers. They are one of the smaller breeds and mature much more slowly so are not ideal for a summer raising routine. Light fleece (6–8 pounds). Mature rams average 175 pounds; ewes, 125.

Dorset: White faces, horned or polled (naturally hornless). Can be bred at any time of the year. Ewes are very prolific (150 percent lamb crop) and good milkers. Fleece averages 8–19 pounds. Rams average 200–225 pounds; ewes, 150–175.

Hampshire: Black faces, ears, and lower legs. The ewes are prolific and good mothers. This is a large, fast-maturing sheep, and for that reason is excellent for summer raising, especially in cooler regions where the pasturing season is short. Fleece averages 7–8 pounds. Rams average 250 pounds; ewes, 180.

---

*Lamb crops are commonly measured in percent as an average of lambs weaned. 100 percent would indicate one lamb per year weaned; 200 percent would mean twins weaned each year. 150 percent is the mark most breeders shoot for.

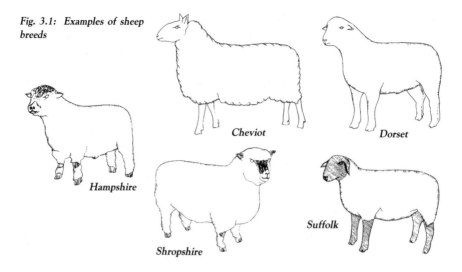

Fig. 3.1: Examples of sheep breeds

Cheviot

Dorset

Hampshire

Suffolk

Shropshire

Montadale: Small heads, open, white faces, clean legs. Montadales are prolific and good mothers. They are a very good dual-purpose breed with heavy (10–12 pounds) fleeces.

Oxford: A large sheep with gray to brown faces, ears, and legs. A fast-maturing breed whose ewes are prolific and good milkers. Rams average 300 pounds; ewes, 175–250. Heavy (10–12 pounds) fleece.

Shropshire: Deep brown or black feet, noses, ears, and legs. Wool covering faces often results in wool blindness, which is a drawback, but breeding is aimed at eliminating this. A good combination wool/meat breed that produces a large carcass and a fleece that averages 9 pounds. Rams weigh 175–200 pounds; ewes, 135–150.

Southdown: The most desirable meat breed. Blocky and short-legged with gray to brown faces. Southdowns are small but mature very early. The fleece is light (5–7 pounds). Rams weigh 175 pounds; ewes, 125.

Suffolk: Open, black faces, black legs and ears. The ewes are good milkers and very prolific (150 percent lamb crop) and have a fleece weighing between 8 and 10 pounds. Rams average 250 pounds; ewes, 180.

Tunis: This is a hornless, open-faced sheep, with a tan or red face. The breed is fairly rare in this country A good milker. Lambs are early-maturing. Rams weigh 150 pounds; ewes, 110–125.

It is not really desirable to purchase purebred sheep, except perhaps in the case of a ram. Purebreds cost more, will not grow faster than good-grade sheep, and will grow more slowly than crossbred animals. As with other animals, they lack the hybrid vigor associated with crossbreeds. As a result, they are not so hardy and have a lower lamb survival rate.

PURCHASE

In choosing a breed to start your flock, you should be influenced by the following factors: personal preference, availability, products desired (wool or meat), and, to a lesser degree, weather and other environmental conditions.

You will do best to choose a breed that is pleasing to you over other factors since it will not be easy to sustain interest in a breed that is unattractive. If you want both wool and meat, one of the medium-wool breeds should suit you. Interest in sheep for wool only and not slaughter would lead one to choose from a breed where wool production is higher. If you have sparse or rocky and hilly pasture, you should consider one of the hardier breeds of sheep. In areas with severe and snowy winters and where undergrowth is thick, closed-faced breeds will be more susceptible to wool blindness.

Avoid livestock auctions in buying your sheep unless you are truly expert at selection. Your best bet is a reliable sheep farmer. A sheep farmer will be better able to understand what you are looking for and should have a greater number of animals to choose from. The farmer can also be very instructive.

You will be looking for all or some of the following: grass lambs to raise over the summer for butchering in the fall, ewes for breeding, or a ram or rams for breeding.

Grass Lambs

Buy your grass lambs as soon as you can after the grass becomes edible in the spring. This will give you a longer pasturing season and more lamb for your money. You will be offered ram lambs or wethers (castrated males). Ewe lambs generally are not sold for meat but are reserved for future breeding and are higher in price. While over a few years a wether will weigh slightly more than a comparable ram, for the purposes of raising over the summer, differences in carcass and taste are nil. If you have breeding stock, however, get wethers or do the castrating yourself, or you will have frequent headaches trying to keep your meat lambs from breeding with your ewes.

Make sure that your lamb's tail is docked, since this ensures cleanliness and helps prevent infection or infestation with maggots.

Look for: Large, well-formed lambs that are straight-backed and stand squarely; a deep chest; good width between front legs which are well muscled and firm; alertness; and lambs that walk with ease. Part the fleece and look for bright pink skin. The mucous membranes around the nose and mouth should be bright red.

Avoid: Generally the opposite of above: small, listless, or generally unhealthy-looking lambs; or those with diarrhea, colds, or runny noses. Avoid poorly proportioned lambs and those with dark, bluish skin or those with crooked legs or which are lame. A sheep whose front legs appear to emerge from one cavity may indicate inbreeding.

Note: If you add grass lambs to your current flock of older breeding ewes and possibly a ram, savage bullying of the lambs may result. We have discovered that adding grass lambs in pairs gives the newcomers some respite and helps alleviate the problem.

Breeding Stock

Ewes: In purchasing your breeding stock, considerably more care is called for. While a grass lamb is raised for perhaps 6 months and slaughtered, a ewe used for breeding hopefully will be around for many years and will affect your entire flock. You can buy a ewe lamb in the spring to raise for breeding or purchase older ones in the late summer or early fall. Especially for the beginner, it is advisable to buy a ewe 2 or 3 years of age that has already lambed. Such a ewe will be mature and have had lambing experience, which will mean fewer demands on the novice owner. If possible, buy one that is already bred so you can start off the first year without having to purchase a ram and worry about a successful breeding.

Avoid livestock auctions and private owners who may try to foist their sickly ewes or poor lambers on you. A sickly ewe can barely support herself much less carry and milk a lamb. We've had good bargains buying culls from a large sheepman. These are not culls in the usual negative sense but normal, healthy sheep from a very large flock that do not compete well for food in large groups. In our small flock they thrive. What is a loss for an owner of a large flock is your gain at a bargain price.

Look for the qualities you want in a good grass lamb. Look also for relatively large ewes because large animals tend to have larger, faster-gaining lambs. Twinning is inherited; therefore, a ewe that is a twin, or has produced twins, is more likely to do so. Avoid shallow-bodied animals or those that are long-legged or with narrow heads. Again, these points are relative, and experience will help you to judge your stock. The fleece should be uniform, compact, bright, and clean; avoid those with uneven coloring. The sheep should walk well and the feet should be well formed and uncracked. Foot problems will affect how well a ewe can carry her lamb. Avoid those with foot rot and those that are limping.

In large flocks, sheep of the medium-wool breeds don't begin to decline physically until they are about 6 years of age. Older sheep cannot feed well when grouped with younger, stronger ones. As explained earlier, older sheep that are culls from larger flocks can be a good addition to your small flock if the price is right. With the proper care given by a conscientious stockowner in a small flock, they can usually lamb for many years after they are "washed up" in the larger flocks.

Avoid young ewes (2–3 years) that haven't lambed yet. If a ewe hasn't lambed successfully by that time, she probably never will. With practice, you will be able to spot such animals. If the ewe's teats are very small (relative, of course, to her age), it is probable that she hasn't had a lamb—or at least one that's lived so that she could nurse it. If there is any doubt in your mind, don't buy.

With all this discussion of age in buying, it would be helpful to know how to determine it. If the sheep is a purebred with papers, the task is quite simple. If there are no papers, the development of the teeth is a fairly accurate, albeit not infallible, indicator of age (Fig. 3.2). A lamb has narrow teeth corresponding to baby teeth in humans. At the age of 12–18 months the centermost teeth are dropped and replaced by two broad permanent teeth. Each year thereafter an additional pair of permanent teeth appears until, at age 4, the animal has what is known as a "full mouth."

After this the determination of age by teeth is less accurate. Often the full set of teeth is intact until the age of 8 or 9, but more often a few teeth are lost around age 5 or 6 resulting in a "broken mouth." Older sheep with no teeth are known as "gummers" and are not (no surprise) in high demand. With experience, the approximate age of older sheep can be determined by the degree of wear or spread of the teeth and the number of teeth lost.

In a sheep with a full or broken mouth, fitness and age can further be determined by the condition and placement of the rear teeth. By feeling to the rear of the mouth (caution: sheep have strong jaws; have another person hold the mouth open while you do it), one can check the condition and shape of the rear teeth. Those with teeth that are still intact and meet well are apt to be of producing age, while those with missing teeth and abscesses should be avoided.

Fig. 3.2: Age determination by stage of teeth development (see text). (From **The Production and Marketing of Sheep in New England,** *a New England Cooperative Extension Publication)*

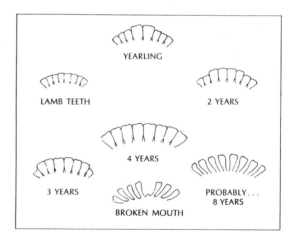

YEARLING

LAMB TEETH

2 YEARS

4 YEARS

3 YEARS

PROBABLY... 8 YEARS

BROKEN MOUTH

Perhaps the most important single consideration in the purchase of a ewe is her udder and teats. A ewe, no matter how many twins she drops, won't be worth much unless she can milk them adequately. The udder should be soft, pliable and free from lumps. Avoid animals that have abscesses or ruptures of the udder and blind or missing teats or those that are abnormally large or thick.

Rams: In choosing a ram, spend even more time, often beginning to look many months in advance so that you'll find a suitable one and not get stuck with a poor substitute. For, while a ewe passes her traits on to one or two lambs per year, your ram will affect every lamb in your flock. In ewes, you should get good grades or crossbreeds, but for a ram, spend a few extra dollars and get a purebred. In this way you'll be constantly upgrading your flock and imparting hybrid vigor. At worst, use a grade or crossbred ram that embodies as many desirable qualities as you can find (rate of gain, size, twinning, etc.). But remember that you won't be upgrading the quality of your sheep. If you breed with a ram that is better than your flock, you will in the long run get a better flock. By breeding to a ram that is equal to or poorer in quality, you will downgrade your flock or at best not improve it.

In addition to the desirable qualities mentioned earlier in this section, select a ram that is large and thick with plenty of bone. Be aware of inherited traits listed in Table 3.2, which is also applicable to ewes. Choose a ram that is "full of it," that is, active and *ram*bunctious. He should be between 1 and 6 years of age, with strong legs and wide at the ears. Also, last but, most certainly, not least, make sure your ram has two fully developed testicles.

Table 3.2: Heritable Characteristics in Sheep

Meat Characteristics	% Heritable	Comments
Twinning	15	Twins increase pounds of lamb sold.
Birth weight	33	Larger lambs at birth usually make faster gains.
Weaning weight	33	Weaning weight is tied to cost of production per head.
Yearling weight	43	This is the ability to gain on own without dam.
Post-weaning daily gain	71	Gain without mother's milk is efficiency of animal to produce on own.
Efficiency of gains	15	Pounds of feed per pound of gain important in economy.
Body type	12	Not highly heritable, but affects purebred value and market value.
Condition score	12	Important in lambs sold at weaning.
Wool Characteristics		
Face covering	43	Ewes with open faces produce 11.1 more pounds of lamb per ewe than closed-faced ewes.
Neck folds	35	A defect in certain breeds.
Grease fleece weight	47	On average, 33% of income from sheep is from shorn wool.
Staple length	45	Major factor in determining fleece weight.
Fiber diameter	57	Important in determining wool grade.

Prepared by Byron E. Colby, Veterinary and Animal Sciences Department, University of Massachusetts; quoted in part from *Genetics of Livestock Management* by John F. Lasley and *The Stockman's Handbook* by M. E. Ensminger; and made available by the Vermont Extension Service in cooperation with the Vermont Sheepherders Association, 1975.

HOUSING

Sheep are perhaps the most defenseless of farm animals, and fencing and housing are as much to keep predators out as to keep the sheep in or to protect them from the weather. If you plan to raise only grass lambs and not winter or breed them and your area is relatively free from stray dogs and other predators, you're in luck. You can get by with tethering your sheep and providing them with shade in warmer weather.

Tethering

In tethering, you attach a chain to a collar on the lamb and to a post with a swivel on it. With a swivel the lamb can move about with a minimal chance of wrapping the chain around the post. I have seen large anchors intended for dogs that screw in flush with the earth. These will serve the same function as a stake but eliminate the biggest danger in tethering: the lamb wrapping itself around the post and choking.

We live on a road with little traffic and no stray dogs, and the first year we had two sheep we were able to leave them loose most of the time. A handful of grain kept them hanging around every morning but they never wandered far and put themselves to bed in the barn every night. This "honor system" worked well until a friend's child chased the lambs one day, and they suddenly realized that they could go anywhere they wanted. From then on tethering was a must. With most sheep you can tether one and the flocking instinct will keep the other in sight. With a larger number of sheep the mathematics become more complex. One summer with 5 sheep, we could tether 3 and keep the other 2 around; but if we tethered only 2, the 3 would wander off. Apparently sheep can add and subtract.

As we discuss sheep wandering off, it is worthwhile to recommend selecting one lamb (perhaps the most rambunctious and thus the most likely to escape) as the "bell lamb." If they wander you can hear the bell at some distance. Try finding a lamb without a bell in the woods and you'll see what I mean.

If you tether them, you can't just leave your lambs unattended on the "back forty" until butchering time in the fall. It works best when they are close to the house and you can see them easily—one of the pleasures of sheep. They must have shaded spots to retreat to during hot weather. Be sure there are no obstacles such as trees, heavy weeds, and briers, in which they may become tangled. If you tie them out during the day and lock them up at night as a precaution against predators, realize that during hot weather they often eat nothing during the day, choosing to graze at night when it is cooler. If you enclose them at night, they will gain slowly or even lose weight. You must be on guard against stray dogs making quick work of a defenseless lamb, rendered even more defenseless by tethering. If dogs are a problem, you'll probably have to go on to more elaborate and expensive methods of containment.

Check the lamb's neck periodically for any sores or chafing from the collar. If not treated, they may become infected and infested with maggots. Most of all, check your lambs daily and unwind their chains so they do not become wrapped around the stake. Tethering is a very good and cheap system for a few sheep *providing you are vigilant.*

Woven Wire Fencing

Sheep fencing is the most disagreeable thing about keeping sheep. It is expensive. I wouldn't even consider fencing large areas unless I were going to breed and winter sheep. We've tethered up to 5 sheep at a time, and that was our limit. It was more than we could safely handle—in fact one hung itself on a tether that summer. At the least you might buy a small roll of woven wire fencing to confine and protect your sheep at night or when you go away. Another option is to move that roll of fencing around as the sheep graze down one area. This is fairly simple and will provide confinement and protection relatively cheaply.

For permanent fencing using woven wire, we have what I term "conventional" and "deluxe" systems. The conventional fence, and one that will do nicely in most instances, is 39- to 48-inch woven wire attached tightly to stakes so dogs can't get under it or over it. If you still have problems with dogs or your sheep get out, the deluxe system, patterned after the USDA design, has 32-inch high 6-inch-stay vertical woven wire fence with one strand of barbed wire running along the bottom and 2-5 strands at the top, giving the fence a total height of 5 feet. This will keep out any dog and most people, and will cost accordingly. If you really want to go into sheep farming, this is the fence for you; but most people will get by with the conventional setup. Young lambs can sometimes slip through the holes in this fencing and may wander off. To prevent this, make them wear a triangular yoke made out of lath. It should measure 8–12 inches on a side. This will prevent them from slipping through the fencing and will not hamper nursing.

Electric Fencing

Most people give sheep very little credit for intelligence, perhaps deservedly, but they can be taught to stay behind an electric fence. This can mean big savings for those who want to fence a large piece of land. It is considerably cheaper than woven wire and easier to install. However, it does require frequent maintenance and checking, and if you live in a snowy region, will require resetting every spring.

For most purposes a two-strand fence of 12- or 14-gauge galvanized wire set at 12–14 inches and 22 inches will be adequate. Young lambs may still get under such a fence so if their wandering becomes a problem, you'll have to go to a 3-strand fence set at 8–10 inches for the first wire, 26 inches for the top, and split the difference for the middle wire. An electric fence will generally keep dogs out and sheep in. The key factor is that the wire must be *very tight* for sheep (more so than for any other animal) or the wire will give and not push through the fleece to the skin. The fence charger should be the finest available and be correctly installed. As with any electric fence, it is necessary to check it periodically to make sure nothing is grounding it, thereby losing its charge.

Sheep should be trained before putting them in this type of fence to make them wary of it and to prevent them from trying to get through it right away, thus getting entangled and shocked and breaking it in their panic. Train in the spring after shearing and ideally after a heavy rain that soaks both the sheep and the ground. Make an enclosure 3–5 feet by 5–10 feet, fenced with electric wire, and set the charger on it. Place a pan of grain in it and let the sheep in and allow them to attempt to get the grain. Watch them closely and be quick to repair the fence if it breaks down or is grounded. Young sheep will learn more quickly than older ones who have never felt an electric fence. This, and occasional jolts from their permanent fencing, should render them fence-wary until the next spring.

Shelter

If you are going to raise only grass lambs, you will have to provide shade for them, but no building or enclosure unless predators are a problem. For wintering sheep, especially in colder climates, some sort of enclosure is required. It can be a 3-sided lean-to open to the south or southeast, or, if early lambing is desired, it should be enclosed and relatively draft-free. For later lambing or in warmer climates, lambing can take place outdoors and a lean-to will suffice. Do not cramp your sheep or force them to live in a filthy, foul-smelling pen. Allow at least 16 square feet of floor space per ewe and furnish good bedding (sawdust, wood chips, straw), and keep the pen clean. Clean pens not only make for healthier sheep, but also make for better-tasting meat. The reason much commercial lamb has that heavy, "lamby" taste is that they are often forced to live in cramped quarters and sleep in their own excrement, which taints the meat.

Door openings should be at least 8 feet wide to prevent sheep from becoming crowded in doorways and injuring themselves and unborn lambs. It is helpful to have electricity in the pen for a heat lamp in case of early lambing when warmth is needed. Lights also make working in the pen, especially during lambing, much easier. Have an outside yard off the sheep barn so that in winter the sheep, especially the ewes, can exercise. We have our barn within the fencing and feed our sheep outside in the barnyard all winter. In this way they get plenty of fresh air and exercise.

Lambing Pen

If your ewes lamb when the sheep are back on pasture, no special equipment is necessary. However, if lambing is to take place inside, you should be able to furnish a lambing pen. This is a portable enclosure that you use to isolate the mother and her lamb(s) from the rest of the flock for the first few days. This permits them to get acquainted and reduces the chances of a mother losing or abandoning her lamb. The pen should be about 5 feet square and must be in a draft-free location. If your barn is large, as ours is, you can build permanent lambing pens. We use ours in the fall to keep our ram (or rams) away from the ewes until we wish them to breed. These pens can also be used to isolate sick sheep from the rest of the flock, to house a veal calf or some weanling pigs—or to confine a neighborhood child who will not quit chasing your sheep.

EQUIPMENT

If you don't have a stream or other running water, you'll have to supply your sheep with a water bucket or a trough. Sheep don't drink as much water as most other stock, and much of their water needs in the summer are supplied by lush grass and dew, but a pail of water kept out of the sun should be available at all times. When sheep are tethered they will wind the chain around the water bucket and overturn it. If you're raising grass lambs, all you'll need is a water bucket and a small dish or discarded pan for feeding grain. For feeding hay and grain in the winter it is best to have a combination hay and grain feeder similar to the one pictured in Fig. 3.3.

The first winter we kept sheep we fed them hay on the ground, and I would guess conservatively that a third of the hay was wasted: urinated on or trampled. Although they do not quite equal goats in their refusal to eat off the ground, sheep will waste a lot when fed that way. Since we began using a manger, there has been almost no wasted hay. Whatever form of manger you use, remember the space between the slats that the head goes through should be 7 inches so the head can fit in but not pull out too much hay at one time. The width of the manger is variable, but be sure to allow 15 inches for each mature sheep and 12 inches per lamb.

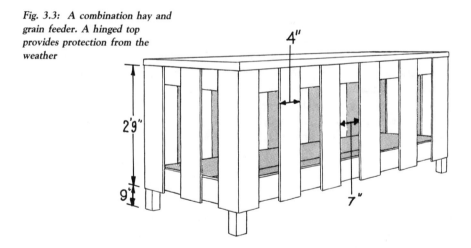

Fig. 3.3: A combination hay and grain feeder. A hinged top provides protection from the weather

A box or dish should be provided, free from tipping and protected from the weather, to supply mineral salt to the sheep. It should be supplied in loose form, since salt blocks may chip teeth and subsequently hamper the ingestion of feed. (At times salt blocks are acceptable: see *Feed* section.)

If you wish to feed your lambs extra grain, which is often recommended, you will need a creep feeder—a feeder or a space that is accessible only to the young animals (Fig. 3.4). The easiest method is to use an existing lambing pen or build one in the corner of your barn and construct the opening so that it is 12 inches from the ground. The lambs can get in at the grain or hay while the sheep cannot.

Fig. 3.4: A simple creep feeder for lambs

FEED

Sheep are desirable because they can grow and produce meat, wool, and lambs by deriving 90–100 percent of their nutritional needs from plentiful (and inexpensive!) pasture and roughages. As with the other multistomached ruminants (cows, goats, and camels), they can take low-protein feeds and convert them into rich food. In the first of four stomachs, the rumen, the food is worked upon by millions of bacteria and protozoa, and low-grade foodstuffs are transformed into essential protein amino acids. Also, many essential vitamins, chiefly the B-complex group, are produced and they need not be added to feed rations.

The rumen is nonfunctional in a young animal, and so for the first few weeks of life, a complete diet consisting of ewe's milk or a good creep feed must be supplied.

Grass Lambs

Rams raised for meat or wool only and to be butchered in the fall are the easiest of all animals for me to feed. I supply them with good pasture, plenty of water, salt, and that's it. As a rule I do not grain any of my grass lambs. I do give them a handful now and then as "candy" to catch them or lead them somewhere, but none other than that. So far I haven't been convinced that giving them extra feed is worth it. Let us say, for example, that a lamb that costs you $30 in the spring will dress out at 30 pounds in the fall with no grain at all. That comes to $1 a pound for your meat. If you feed that lamb 100 pounds of grain concentrate over its lifetime (at $8 per hundredweight) and you realize a gain of 5 pounds, your meat then cost $38 for 35 pounds of meat or almost $1.09 a pound. Even if you add 10 pounds to the weight, which is unlikely, your price will drop to 95 cents a pound–a saving that is not all that remarkable. But experiment for yourself. With the availability of supplemental feed in your region, you might be able to make it worthwhile. Consider, however, that if you breed your own lambs and they cost you $10 apiece, $8 (or 100 pounds of grain) will almost double their cost.

If your lambs make slow gains or seem to be lagging, it could be that your pasture is inferior. In such a case, and in the fall when the pasture is dying, grain will be called for in the diet. I rarely feed more than 1 pound a day.

Pasture: A good pasture can handle up to a dozen ewes and their lambs per acre. This is a maximum; most pastures will furnish far less grazing. Watch your sheep carefully and note the condition and length of the grass. Sheep will graze very closely, if allowed, and even kill pasture, so watch them to prevent overgrazing. If you don't have additional pasture, you may have to give up a few sheep (or tether them on your front lawn) until you achieve a balance.

The best system involves at least two pastures so that when one gets eaten low you can shift to the other and allow the first to regrow. This rotation of pastures is also important in the control of parasites. Sheep will improve a pasture in the long run, weeding, fertilizing, and spiking in the fertilizer with their cloven hooves. A pasture that handles only 4 sheep per acre may handle 6 in a couple of years. It is also important to keep your pasture in shape with lime and additional fertilizing when needed. The best sheep pastures include alfalfa-brome grass, clovers, clovers and grasses, ladino clover mixed with other legumes and grasses, and orchard grasses. Another good source of grazing is cut-over hayland after baling.

Watch for plants that are poisonous to sheep, such as lupine and choke cherry. Usually these plants will not be eaten unless pasture is very low and there is no other suitable forage available, but *veratrum* (false hellebore) is relished by sheep and can cause birth defects if eaten later in pregnancy. Burdocks can become hopelessly entangled in wool and make it worthless.

Water and salt must always be available to sheep. While loose salt is preferred over salt blocks because of less danger of injury to teeth, I have found that some of our sheep dislike the consistency of loose salt and prefer the salt lick. It is better to give it in that form rather than not have them get any. Two of the most important minerals for ewes are calcium and phosphorus, essential for the growing lamb and the production of milk. Often these minerals are missing from pasture and hays, especially if poor quality hay is used, and must be supplied as a supplement. Always supply free choice in weatherproof feeders a mixture of either $1/2$ mineral salt and $1/2$ dicalcium phosphate or $1/3$ mineral salt and $2/3$ steamed bone meal.

Important vitamin needs for sheep are A, B complex, D, and E. If you are feeding good hay with at least $1/3$ legume, your vitamin A needs should be met satisfactorily. Commercial grains usually contain vitamin supplements, so check the bag label if you fear there are deficiencies. Other sources of vitamin A are cabbages, corn, squash, and carrots, which are all relished by sheep. As long as your sheep receive at least a few hours of sunlight a day their need for vitamin D should be met. Vitamin E deficiency may evidence itself by stiff-legged lambs. If this is a problem, wheat germ oils should be fed and a veterinarian should inject your lambs with vitamin E. B-complex vitamins present no problems because they are synthesized in the rumen of sheep.

Hay: Hays of the type listed for pasturing will be suitable. If a high-quality legume hay or legume-grass mixture is used, it is possible to save substantially on feed bills by eliminating grain (see Tables 3.2 and 3.3). Second- or third-cutting hay, while usually a bit more expensive, is well worth it because sheep relish this finer hay. Because fewer hard stems are present, less is wasted. Do not force sheep to eat the coarser hay by leaving it in the manger until they clean it up. It is less nutritious and may be removed for bedding or fed to pigs or other less-choosy stock. We generally feed the remnants to our horses. A mature sheep will consume 500–700 pounds of hay in a winter.

Don't feed sheep a feed concentrate made up solely of corn. Also, don't put sheep out on a very rich clover or alfalfa pasture that is wet with rain or dew, or stomach bloat may result. As with all stock, feed on a regular schedule.

Ewes

Proper feeding of ewes will lessen stillborn births and result in healthier lambs that will gain faster. It will prolong the productive life of the ewe, increase her milk flow, and lessen the chances that she won't accept her lamb because of her weakened condition.

As with other stock, flushing (improving the ewes' diet) will improve their breeding record. Flushing two weeks before and up until breeding will tend to bring all the ewes in heat at about the same time and hence give you more uniform lambing times. It has been shown that flushing will increase your lamb crop by about 18 percent. Ideally, you will have a lean ewe gaining weight at the time of breeding. You do not want your ewes to be too fat or the result may be many unbred sheep. If your ewes are overweight, switch them to sparser pasture 6 weeks prior to flushing so that when they are ready to be flushed, they will be down to normal weight. Flushing can be accomplished by switching the ewes from fair to excellent pasture, or if on good pasture or already on hay, adding $1/2$ to 1 pound of grain to each ewe's feed per day. Upon successful breeding, feed as indicated in Tables 3.3 and 3.4.

It is just as important not to overfeed your ewes as it is to make sure they aren't underfed. Above all, ewes need plenty of exercise during pregnancy. Feeding them outside will make them exercise and also expose them to healthful sunshine.

Sheep need a grain that is 14–16 percent protein. The next most important factor is its palatability to the flock. The generic names for these feeds may be sheep feed, goat feed, or horse feed. These are usually a balanced ration, containing vitamins, minerals, and other necessary foodstuffs. We have found 14–16 percent dairy ration to be both a relatively cheap feed and a very palatable one for our sheep. Experiment with different feeds, and find the one most favored by your sheep and your pocketbook. Good high-protein supplements include soybean, cottonseed, and linseed oil meals.

Use what feeds you have to make up the most economical mix. Once you choose a program, try to stick to it; if you must change, do so very gradually. Be sure that silages are not spoiled or you may lose some sheep.

Table 3.3: Suggested Daily Rations for Ewes in Good Health from Breeding until 6 Weeks Prior to Lambing

#1	Grazing on good pastures may be continued, but care should be taken not to overgraze legume seedlings or they will be winter-killed.	
#2	Barn feed legume hay	3–4 lb.
#3	Grass hay	3 lb.
	32–36% protein supplement	$^1/_2$ lb.
	or	
	Commercial mixed grain (12–16% protein)	1 lb.
#4	Mixed legume and grass hay	$2^1/_4$–$3^1/_4$ lb.
	Protein supplement	$^1/_4$ lb.
	or	
	Mixed grain	$^1/_2$ lb.
#5	Legume or grass-legume hay—good quality	$1^1/_2$–$2^1/_2$ lb.
	Corn silage	4–6 lb.
#6	Corn or nonlegume silage	8–10 lb.
	High-protein supplement	$^1/_4$–$^1/_2$ lb.
#7	Good grass or legume silage	8–10 lb.
	Hay	1 lb.
#8	Late cut timothy hay	3 lb.
	Corn	$^1/_2$ lb.
	Oats	$^1/_2$ lb.
	Bran	$^1/_4$ lb.
#9	Grass hay	3–4 lb.
	Mixed grain (14–16%)	1 lb.

Up to 4 weeks before lambing you can save substantially on feed if you have your own legume hay. The less nutritious your hay, the more you will have to pay for extra feeds. You can also save by not overfeeding your sheep. Up until a month before lambing the sheep should be well covered over the backbone and ribs, but you should still be able to feel bone in these parts. If you can't feel the bone, you're overfeeding. As Table 3.2 indicates, you should increase feed a month or 6 weeks before lambing because 70 percent of the growth of the fetus takes place during this time. Follow recommendations. It is just as bad to overfeed during this time as any other. If overfed, the ewe will be too fat, the fetus will weigh too much, and you will have lambing difficulties and possible loss of lambs.

When signs of lambing are imminent, cut back on bulky feed and add $^1/_2$ pound bran to the ration as a laxative so there will be less competition for room between the end of the large intestine and the vagina at lambing time. This is especially important for first-year lambers and ewes that have a history of difficult lambing. After lambing, don't grain for 24 hours, since udder problems may occur and the

Table 3.4: Suggested Daily Rations for Ewes 4–6 Weeks Before and up to Lambing

#1	Legume hay	3–4 lb.
	Oats, bran, corn or barley (or mixture of all in equal parts by weight)	³/₄ lb.
#2	Grass-legume hay	1¹/₂ lb.
	Corn silage	4–6 lb.
	Mixed grain or bran and oats	³/₄ lb.
	High protein supplement	¹/₄ lb.
*#3	Timothy hay	3 lb.
	High-protein supplement	¹/₂ lb.
	Mixture of corn, oats and bran or commercially mixed grain	³/₄ lb.
#4	Mixed legume hay	3–4 lb.
	Mixture of oats and bran or bran and corn or commercially mixed feed	1 lb.
*#5	Corn or nonlegume silage	6–8 lb.
	Good nonlegume hay	1 lb.
	Protein supplement	¹/₄–¹/₂ lb.
	Mixture of bran, corn, oats or oats and bran or commercial mixed feed	³/₄ lb.
#6	Legume hay	2 lb.
	Timothy hay	2 lb.
	Mixture of bran and oats, or bran and corn	1 lb.

*These rations may be too low in calcium. In this case or anytime nonleguminous roughages are fed, feed a mineral supplement as suggested earlier. These tables are from *The Production and Marketing of Sheep in New England,* a New England Cooperative Extension publication.

lamb may scour. Then gradually build the grain back up to 2–3 pounds per day depending on how the ewe milks and the lamb reacts. In the case of twins, which are quite a strain on the ewe with their increasing demand for milk, adequate grain is critical. Other feed such as carrots, beets, cabbage, or cull potatoes fed chopped and not exceeding 3 pounds a day are relished and are an excellent aid in the production of milk. When lambing occurs soon before or after, we put out sheep to good pasture. We give up to 3 pounds a day of mixed grain in addition to the pasture and taper it down to nothing in 6 weeks or upon weaning, depending upon the condition of the lamb(s) and ewe.

Creep feeding is essential to ensure that lambs become established on a diet of high-quality hay and grain as quickly as possible to bring them to market at maximum market weight rapidly. Early in their lives lambs are not true ruminants, but by providing early access to roughage and grain, they are afforded the opportunity to develop this characteristic as rapidly as Mother Nature will permit. Baby lambs should not have to compete with their mothers and other ewes for roughage as they are sure to come up short at the very time they need the best feed possible in an unrestricted manner.

Breeding Rams

Generally, a ram will get by quite well on 3–4 pounds of good hay per day. If he seems weak or thin, give him a little grain—but he probably has worms or some other health problem. For 2 weeks before and during the breeding season I give him a $^1/_4$ to $^1/_2$ pound of grain per day to provide the additional nutrients he needs during this period of increased activity.

Replacement Ewes and First-Year Lambers

I feed our replacement ewes a pound of mixed grain per day all summer into the breeding season to give them the extra feed needed for faster growth and the rigors of lambing. Follow any of the programs outlined in Table 3.2 for them, but supply one pound of mixed grain (or the equivalent) extra for growth. Your first-year lambers are growing themselves as well as feeding a fetus. Watch to see if they become too fat or rundown and adjust their feed accordingly.

The problem of feeding different rations to your sheep in the winter (e.g., young ewes needing more grain than older ones; ewes needing grain but rams not) is solved by large operators by putting sheep into different winter quarters according to what they eat. The small sheep owner wouldn't want the expense of different pens nor should he have to go to the bother with a small flock. Neither would he want to feed a ram 2 pounds of grain a day when he doesn't need it, or a ewe more than she needs because the younger ewes need that much. It is both expensive and possibly injurious to your ewes to overfeed. We have handled the problem by herding all our sheep into the barn and closing the gate. We then herd out those who get one ration, and keep in those that get another. This is difficult at first, as *all* the sheep want to go out. But do this for a week and your "dumb" sheep will learn. Herd them in, then open the gate, and the ones that usually go out will do so, while those that are fed inside will stay in.

Supplementing Commercial Feed

Because of the "protein factory" in the stomach of ruminants, sheep can be fed cheaply on conventional feed alone. While grass lambs can be raised to butchering weight on grass alone, breeding ewes will generally need other food besides grass or good-quality hay. The numerous feeding programs outlined earlier should afford stockowners a greater opportunity to feed their flock on what they can grow themselves, or what they can purchase cheaply in their locale.

As mentioned earlier, cull potatoes, cabbage, turnips, and carrots are a good supplement to regular feed and are especially good for milking ewes. In feeding these items, do not exceed 3 pounds per animal. Do not make a change in the amount of grain or other concentrate you feed, but allow your savings to come in decreased hay consumption.

Leftover bakery products from your supermarket can provide you with a very low-cost protein supplement, but careful dietary management is critical to maintaining a properly functioning rumen. Any ruminant is basically a roughage eater and what we add to the diet is done to speed up the growth of the animal to a desirable market weight, increase the production of milk, or improve reproductive efficiency. Without proper and adequate roughage, a diet slanted too much toward stale bakery products can lead to serious digestive disturbances.

MANAGEMENT

Routines

The routines you establish with your sheep will, of course, depend upon your demands and the size of your flock. If you are simply raising grass lambs, you will butcher them all in the fall and be done with sheep until the following spring when you buy a few more woolly little lambs.

We have a ram and a number of ewes. After lambing in the spring, we keep a few meat lambs and any replacement ewes we might need and sell the rest. This income more than pays for the feed for the ewes the past winter. Any surplus goes to lower the overall price of our meat and wool.

Auctions are a quick way to sell surplus sheep or lambs, but generally a higher price can be fetched through private sales. You can usually expect to get 30 to 50 percent more for ewe lambs because of their breeding potential, but check average prices in your area in a market bulletin before selling. Any runts you may have on hand around Christmas and Easter can often be sold at a good profit (quite often higher than larger full-grown sheep) because of the demands then by certain ethnic markets.

Many sheep owners today gear their lambing time to the Easter lamb market. Develop your fall breeding program to have the majority of the lamb crop ready for that market. You need to program the arrival of your lambs so that they will reach market weight at Easter. If you are located within reasonable driving distance of a major metropolitan area it is entirely feasible to develop an excellent market for your crop. Easter lamb traditionally commands the highest price of the year. Once you have established a clientele, it is possible that you will have a difficult time keeping up with the demand.

A day taken to visit small meat markets in either predominately Italian or Greek neighborhoods may put you in touch with an excellent potential outlet for your lambs.

Another bonus in sheepraising is your wool and hides. Shearing is carried out in late spring or early summer each year. Depending upon the breed, you may get 18 pounds of wool per sheep, but for the medium-wool breeds the average is 5 to 12. Your grass lambs can be sheared before butchering in the fall and the hides tanned (see Appendix D); or, if you prefer, you can tan the hides with the wool intact for use in making slippers, gloves, rugs, and coats. Unless you plan to be snowed in all winter or you have a large family and can delegate work to your children, you probably won't be able to spin all your wool. There are a number of establishments that do custom spinning and dyeing wool. You can send your wool and you get the same amount back (minus a certain percentage depending on how dirty or full of burrs the wool is) dyed to your specifications. All this for 50 to 75 cents a skein! The wool from one lamb will keep you in socks for years.

We generally send all our wool to be spun, and sell it by the skein. We charge far less than the store price for 100 percent wool but still a good deal higher than our cost. This helps to pay off the flock and possibly generate a little profit.

Some handspinners in your area might well be interested in buying your whole fleeces. Also, look into wool pools in your region. These are central points where you take your wool, tagged and bagged. When the wool is sold, you receive payments minus the cost of running the pool.

Keep signed receipts from the purchasers of the wool and at the end of the year file these at your local U.S. Agricultural Stabilization and Conservation Service (ASCS) office so you can participate in the wool incentive payment program. This program is financed out of tariffs on imported wool (not our tax dollars) and pays the seller a premium depending on the wool he has sold and the national average price for wool in this program. The ASCS office near you can give you full details.

Sheep can be beneficial for pasture improvement. As long as you don't allow them to overgraze, they will improve your land. Besides producing rich fertilizer, they also eat about 90 percent of all weeds, so after a few years you should have a lusher, weed-free pasture. Liming and top dressing with poultry or other available or cheap manure should also be done periodically.

Handling

Your problems in handling, or retrieving loose sheep, will be minimized if you tame them early in life or soon after you buy them. One of their greatest defenses is their timidity, and they will shy away from almost anything unless given reason not to. If you feed them by hand with grain, petting them and talking to them often as you do so, they will begin to lose their natural timidity. While this can be helpful if any get away or during shearing, it can also be a great help during lambing if your ewes feel secure and unafraid around you.

If your lambs have escaped their pens, a little grain in a dish will usually lure them back. Do not try to reason with them, they will not listen. In catching especially shy sheep we have found making a snag out of baling twine to be very effective. Make a slip knot and attach 10 or 20 feet of twine to it and sprinkle some grain around the loop. Be patient. When the sheep steps into the loop give a yank and you'll have your stray.

In guiding stubborn sheep to a desired point, grab with one hand under the jaw (gently) and push with the other from behind. Steer with the head. If you really want to get fancy, you can buy a shepherd's staff from one of the farm supply companies.

Unless you have only a few sheep and you can tell them apart, you will need to mark your flock for identification. This helps immeasurably around lambing time and in your record keeping. A commercial ear tagger can be purchased for easy marking, but with a relatively small flock you can do just as well with a concocted system such as a bell on one, a collar without a bell on another, and no collar on another.

Predators

Dogs are by far the most common predators of sheep. Depending upon your part of the country, coyotes too may be a threat. While your sheep are at the mercy of a predator while tethered, a good fence should prevent animals from killing all or part of your flock. Keep an eye out for any strange dogs in your neighborhood, since they are candidates for sheep killing. Be especially watchful when friends come to visit with their dog. Dogs that have never seen sheep before are naturally quite curious. It usually begins quite innocently ("They're just playing" are famous last words) with the dog playfully running after the sheep. Then he gets all worked up, nips, and nips again, draws blood . . . and you know the rest. Prevention is your only recourse.

If you lose any sheep to dogs, local dog license funds usually provide for compensation for the owner. On rare occasions, sheep may be lost to bears or other wild animals. Many states provide reimbursement from fish and game funds, providing you can prove loss by producing the carcass (which is often quite a trick because that is presumably what the bear came for in the first place).

Shearing

Plan to shear in the spring after the weather has settled. There should be no danger of severe weather that would chill the sheep. Wait until there have been enough warm days to bring out the oil in the fleeces. This oil strengthens the wool and also makes the wool easier to clip. Try to shear before setting your sheep to pasture, as new pasture tends to loosen the bowels, which results in stained fleeces. Proper tagging will, however, cut down on staining if you must put them out.

You should remove all stained pieces and burrs (if possible) before shearing. You'll see that it's wise to dig up any burdocks in your pasture because a sheep covered with burrs will have a worthless fleece.

Watch someone shear and then try it yourself. You can use either electric clippers or hand clippers. It is important to remove the fleece in one piece, shearing close to the body but avoiding nicking, and avoiding any second cuts. Do not go over the sheep again as short fibers result and the value of the fleece is lower.

If you do not wish to shear your sheep yourself, there are often traveling shearers who will do the job. Often they will not come by for one or two sheep. If your flock is small, arrange for neighbors to gather their sheep with yours in one place.

Meetings

In most states and areas where sheep are raised, associations of breeders hold informal, periodic meetings. Beginners especially will find these useful—and fun. You can usually watch demonstrations of shearing, docking, or castrating, and the almost incredible herding performance of trained sheep dogs, notably the Border Collie. People often take two or three sheep to be sheared on the spot for a nominal fee. Different breeds may be on exhibit; extension service representatives or veterinarians give lectures and hand out helpful informational circulars or bulletins. Frequently, there will also be spinners and weavers at work, and people from your locale or region who will buy or trade your fleeces for skeins. These field days are enjoyable opportunities to compare notes and problems with large- and small-scale sheep raisers—or with other backyarders.

BREEDING

If you go into breeding, start slowly. It's a good idea to buy a bred ewe or two the first fall or winter. This frees you from the need of a ram and concern as to whether conception has taken place. You might be able to get a good ewe, somewhat advanced in age and not a good competitor in a large flock, but which will do well in your flock. We bought our first ewe in January for $30. She was perhaps 6 years old and doing poorly (although she was by no means sickly) in the large flock; no sooner was she in ours than she quickly established herself as the matron of the flock. To this day she dominates all the other sheep. She has a few more years of lambing, and she has already supplied us with lambs and more than paid for herself.

Most of the medium-wool breeds come into heat fairly regularly, beginning in the late summer or early fall and continuing (if not bred) until January. The heat period is when the ewes will permit the rams to mate. Its duration is 3 to 73 hours but three-quarters of the sheep stay in heat from 21 to 39 hours. The period will occur every 13 to 19 days (with the average being every $16\frac{1}{2}$ days) until the ewe is bred or until the breeding season is over. The Dorset and Tunis and their grades or crosses can breed any time of the year, and caution should be exercised so that they are not bred at inconvenient times. (Inconvenient, that is, for you!)

The ram should be allowed to run with the rest of the flock the entire year except before the ewes begin to come into heat in the late summer or early fall. It's a good idea to be extra careful and not risk early pregnancies by removing the ram in plenty of time. In Vermont, we pull our ram in late August and have never had any "mistakes." One ram from yearling age up until 5 years of age can handle 25 to 30 ewes. During breeding, feed the ram one pound of grain per day. After 6 years, rams begin to lose potency. If you find yourself with a large number of unbred ewes, it may be time for a change.

The gestation period for sheep is 144–152 days with an average of 147. Therefore, about five months or so before you want your lambs, you should move your ram in with the flock. Before doing this, carry out the following procedures: eying (ewes and ram), tagging (ewes), and ringing (ram). Eying involves clipping the wool from around the face and eyes in closed-faced breeds. This improves the eyesight and prevents eye irritation from the wool and seeds and dirt carried in it. Tagging is clipping the wool around the dock of the ewes. This promotes cleanliness and permits the ram to couple with the ewes more easily. Some people shear a ram prior to the breeding season, but in a cold climate such as Vermont we do not do this because the ram will not be able to grow a full fleece back before the onslaught of winter.

Ringing is an acceptable substitute and consists of clipping wool from the neck and from the belly around the penis. This makes it easier for the ram to make proper contact with the ewe during mating. Excessive heat may affect the potency of your ram adversely, and if breeding is to be carried out in very hot months, complete shearing may be necessary to ensure full potency. Your sheep should not be wormed during the month before they are bred.

Letting your ram run with the flock all year (with the noted exceptions) is good because the ewes will know him and therefore be more likely to accept him. If possible, when you separate the ram, try not to alter his feeding or grazing routine very much, if at all, since this could affect his performance. Two weeks before setting your ram in with your ewes, you can attempt to flush them. Secure your ram in sight of the ewes (but still out of contact). This is called "teasing" and can help stimulate ovulation in the ewes.

It is very difficult to tell when ewes are in heat. Often they will begin mounting each other and their vulvas may be red and swollen, but this swelling is not nearly as pronounced as in pigs. You can usually tell if they allow the ram to mount them. In most cases, a ewe will not be receptive to the ram unless she is in heat. If you can see your ram breed with your ewes, you calculate approximate lambing time from that date—*assuming* it was a successful insemination. Or you can just leave the ram in and assume all your ewes are bred, and have the lambs come as a surprise the following spring.

A better system, in which you can be reasonably sure of the date of lambing, is to equip the ram with a marker on his chest so he marks the back of the ewe when he mounts her. You can buy a harness with a crayon attached from farm stores or you can mix a marking paste to apply to the ram's chest once a week or so. A simple paste can be made by mixing mineral or linseed oil (do *not* use motor oil) and any available nontoxic dye such as venetian red or lampblack. Change the color of the crayon or paste every 16 days, using in progression yellow, red, and then blue or black. They are used in this order because they cover the previous color.

The procedure: Equip your ram with the first color. Within 16 days (the average heat cycle) all the ewes will come into heat and should be mounted by the ram. They will then be marked by the first color. The day you notice the mark on the ewe is approximately 21 weeks from their lambing time—*tentatively*. After the 16 days, change to the next color. If the ewes are not marked with this color, it means they are already bred. Those ewes that are bred will not come back into heat, and they will not accept the ram. These ewes should lamb 21 weeks from the day of the first marking. Any ewes marked with the second color should lamb 21 weeks from that day. The color will be changed again after that 16-day period, and so on. In small flocks, all ewes should be bred in their first or second heat after being with the ram. If none of the ewes is bred, your problem is most assuredly the ram. If a few ewes are not bred, they may be too old, in poor health, or sterile. They should be considered for mutton or for sale.

While sheep in commercial flocks are bred for only 5 years, those in small flocks generally receive such good care that they can lamb successfully for as long as 10 years. As a sheep gets older, its teeth deteriorate, hampering food intake, and in such a case it may not breed successfully, or may not be able to nurse the lamb, or it may be so physically taxed that it will die during pregnancy. Keep an eye out for a ewe having trouble. Sell, butcher, or keep her as a pet, living out her twilight years without the drain of children (but still supplying wool, of course).

There is some disagreement as to how young you can begin breeding a ewe. Some say they shouldn't be bred until their second year; others recommend breeding in their first year, to lamb at the age of 14 months or so. Recent studies have shown that ewes bred as lambs will produce a greater lamb crop during their breeding life than those bred as yearlings. Ewe lambs should be at least 9 months of age and well grown if they are to be bred.

Ewes that are bred at an early age must receive a diet adequate to meet the needs of the developing lamb as well as her own needs for growth and maintenance. These ewes must have particular attention at lambing as they are more inclined to have difficulties than older sheep.

These young ewes also are more inclined to refuse to accept their lambs, thus presenting the owner with the additional problem of possibly bottle-feeding the young.

A week or two before you believe lambs are due, check to ensure that the ewe's udder and teats are free from wool. If not, clip that area and around the eyes (tagging again) if necessary.

On the day of lambing, a ewe *may* exhibit one or more of the following signs: She will be off her feed, restless, isolated from the rest of the flock, or dig or paw at the ground or bedding while circling a particular spot. She will appear sunken in front of the hips; the vulva may be red and enlarged; and the udder will be substantially larger and the teats coated with a waxlike film.

When you notice any sign of labor, watch but leave her alone and don't upset her by constantly peeking and feeling her stomach. Sheep may need help at times, but they are not nearly so problem-prone as they are reputed to be. First-lambers tend to have more problems, but if your flock has been well fed and exercised, you shouldn't have any major problems. A friend of ours who has the largest flock in Vermont had 300 lambings one spring and had to assist only in one.

Lambing: What to Have Handy

A clean pail
Heat lamp
Vaseline, mineral oil
Soap
Soft, heavy string
A lambing loop (nylon twine)
Uterine capsules
Dry rags, towels
Ammonia or smelling salts
A helper

Do not leave pails of water where lambing is to take place because it is possible that lambs will be dropped into a pail and drown. Offer water to the ewes periodically and then remove the buckets.

The normal position at birth is the backbone of the lamb toward the back of the ewe, nose presented first with front feet presented alongside it. If afterbirth is passed soon after the birth, generally there will be no more lambs. If afterbirth doesn't come and the ewe continues to show signs of labor, paying little attention to the lamb, another will probably follow. In some instances, 3 lambs (or even more) may be born, but you needn't bother yourself about such occurrences.

If the lambs are born out on pasture, you needn't worry as long as the lamb(s) seem healthy and the mothers clean them off and allow them to nurse. If born inside, cut the umbilical cord off at 4 inches and apply 7 percent tincture of iodine to prevent infection.

If conditions are crowded, place the ewe and her lamb in a lambing pen for a couple of days. You can put the ewe there before lambing if she is obviously ready to drop the lamb. Make sure she can see the rest of the flock and doesn't get upset. This allows the ewe and her lamb to get acquainted without danger of losing each other—most important for yearling and ewe lambs.

In moving the new lamb to the pen, grab it gently under the chest and back it into the pen, keeping it under the ewe's nose. For at least the first few days the ewe knows her lamb by smell only, and if she or it wanders away and it mixes with other sheep, she may disown it.

In chilly weather some burlap or blankets hung around the outside of the pen will help keep out drafts. A heat lamp should be hung in the corner of the pen so that the lamb can get close to it by choice. To prevent ewe or lamb from getting burned, hang it 3 feet above the floor. Do not use it for more than 3 days, because the lamb might get chilled when removed from the pen. If all goes well, the lamb and ewe can be let out of the pen in 2 days and allowed to run with the flock. Follow the feeding instructions listed earlier and you will minimize your problems. You can wean the lamb as early as 4 weeks or let it wean itself naturally. We generally keep all our sheep together and in time the ewe will dry up and not allow the persistent lamb to nurse. As the ewe dries up, or as an aid to help her, slowly taper off her grain supplement until her feed routine is back to normal.

Problems During Lambing

Sometimes 3 to 4 hours are required during lambing. Examine the ewe a half-hour after the first water bag is passed and note her progress. Have all those helping wash their hands thoroughly. If you enter the ewe with your hand, coat it first with a lubricant such as Vaseline. Make sure your fingernails are short and smooth, and if possible bend your fingers at the knuckle before inserting them to avoid scratching. Lay the ewe on her right side and probe with your hand, trying to determine if the lamb is in the proper position. Ideally, its nose should be between its feet. (You can see now the benefit of having your ewes tame and used to your presence.)

A few hints: To distinguish between the front and hind leg, feel above the knee. The hind leg will have a prominent tendon; the front leg, muscle. If the hooves of the lamb are pointed up (toward the backbone of the ewe), they are probably front legs coming in the normal position. If they are pointed down, they are probably the hind legs and the lamb is being presented backwards.

The most common difficulty is when *one or both* front feet are folded underneath the lamb. In the case of one foot folded back, tie a soft string around the presented leg (so it will not be "lost" in this procedure) with plenty of string protruding. Then push the lamb back slowly and gently. By sliding your hand down the neck you can locate the leg that is turned back and straighten it out. The lamb is then in the normal delivery position and should be delivered without further problems. If the ewe is too fatigued and cannot at this time pass it unaided, use nylon twine to make a noose that is placed behind the ears and then into the mouth. String tied around the feet can be pulled along with the loop and the lamb should come. You can pull quite hard, surprisingly, without hurting the lamb. However, always pull *with* the ewe's contractions.

In the case of *both front legs doubled back*, try the same procedure as outlined above. If that does not work, place a loop of string around the lamb's head with the knot in its mouth. Elevate the ewe's hindquarters and gently force the lamb back until you can run your hand down the neck and flip the legs through the pelvic arch. If the head has not slipped back through the arch, start it through before you start the legs through. In all of the above operations, lubricant applied to the vaginal wall will make it easier to maneuver the lamb into position.

Often in the case of hard labor, the head of the lamb swells and makes lambing even more difficult. If the head is turned to either the right or the left and you find it impossible to keep it in position to work it into the birth canal, fashion a loop of twine or nylon string as suggested above. This, then, can be used to move the lamb along.

In the breech position, the lamb is backward, and the rear legs are under the belly of the lamb and pointing toward the front legs. The tail is often the only thing that can be identified. On examining the lamb, the rear legs sometimes are located only with difficulty, so slow, careful, systematic exploration is critical. Once the hocks are identified, they are grasped, one at a time, with the thumb and forefinger.

Slow, gentle traction is applied as the leg is brought backward. If you are successful in accomplishing this maneuver, it is usually possible to slide your hand downward and forward until you can grasp the hoof. With the hoof in your hand, you can turn it inward and, at the same time pull it backward. The same procedure is repeated on the opposite leg and the lamb is then delivered by applying firm traction.

Sometimes you will be presented with an abnormally large lamb. If it is in the normal position, work the legs forward a bit and pull them gently while using the other hand to pull the head from side to side with the fingers positioned behind the ears. A helper can stretch the top of the vulva to ease the operation.

When you are faced with an obstetrical procedure, do not hesitate to call for professional help if you feel you are in over your head. Don't belabor a difficult delivery; instead, recognize your own limitations and call, before the situation gets out of control. A dead lamb is a poor reward for failing to recognize a situation that is beyond your abilities.

Problems After Lambing

The lamb, once delivered, often needs assistance in starting to breathe. Time is of the essence, and the necessary supplies should be nearby in the lambing pen. An old, clean facecloth or piece of towel works well to remove the excess mucus from the nostrils and the rear of the mouth.

The next step should be to give the youngster a brisk rubdown with a towel, especially over the chest, drying as well as attempting to stimulate contractions of the ribs. Don't be too rough—or too tender.

Next, grasp the newborn by the hind legs, elevate it so the front hooves are just off the ground, and swing the lamb back and forth in a gentle arc. This will help to ensure that mucus and other fluids in the back of the mouth and in the nostrils are expelled. Oftentimes this is enough to stimulate the newborn's breathing.

If breathing has not started, lay the lamb on the ground and pour a small amount of cold water into one of its ears. This simple procedure will often stimulate the breathing reflex when other attempts have failed.

Another source of stimulation is to have either household ammonia or smelling salts on hand. Pour a small amount of either solution on a piece of cotton and hold it close to the lamb's nostrils. This can have a powerful effect on stimulating breathing.

If a lamb is chilled, it should be immersed in a pail of water heated to 90–100°F. Leave all but its head in the water for 10 to 15 minutes, then towel it off quickly and place it under a heat lamp. The danger of this method is that much of the lamb's odor will be washed off and the ewe may disown it.

It's essential that the lamb get an adequate amount of colostrum within a short time after birth. Colostrum supplies the critical antibodies that ensure that the lamb will be protected against the many diseases to which it is susceptible in the first days of its life. If these antibodies are not supplied in the first 12 to 18 hours of life, additional colostrum later will be of little or no value for immunizing.

The reason for this is a process called closure, whereby the intestine can absorb the colostral antibodies for only that short period of time after birth.

Nursing

If the lamb does not nurse within an hour, be patient. Allow the lamb to gain its strength slowly. You may have to aid it. Put some milk on the end of the teat and push the lamb's mouth to it or strip some milk into its mouth. Tickling the underside of the lamb's tail while it has the teat in its mouth may help.

If a mother doesn't clean her lamb off, place it by her head to encourage her. If she still doesn't dry it yourself with a clean rag or towel. If she won't accept it or let it nurse, you will have to assist the lamb. Often the reason a ewe disowns a lamb is that she is a first-lamber, is in a rundown condition, or has udder problems. In the latter case, pressure and pain in her udder may make it uncomfortable to have her lamb nurse. Withholding feed the day after lambing will go a long way toward lessening such problems. Try to strip some milk out of the ewe to reduce the pressure. Apply hot packs and keep the lamb nursing if she will allow it.

If she still refuses the lamb, then you can only counter with perseverance and patience. Don't give up. On one occasion we worked for 5 days with a ewe before she finally allowed the lamb to nurse. You must try to get some colostrum into the lamb, either by holding the ewe and letting him nurse or by stripping some from her teat and giving it to him with a dropper. Rubbing some afterbirth or some of the ewe's milk on him will often encourage her to accept him.

A little grain sprinkled on the lamb may make her more receptive. Something to distract and worry her, such as a dog tied within her view, will get her mind off her resistance to nursing. Hold her on her back and let the lamb milk, or face her into a corner so she can't move and hold her back legs so she can't kick him away.

Keep at it. Remember, each rebuff discourages the lamb even more; however, each successful nursing heightens the lamb's confidence and begins to make the ewe more accepting. The thrill of seeing a ewe accept a lamb after 5 days is well worth the effort. If these measures fail, give her another chance the next year. Should she repeat the performance, cull her.

Feeding the Young

[*Editors' note.* Sometimes the lamb refuses the mother. The author of this book happened to be away when one of his older ewes lambed. Since one of our own pregnant ewes was being boarded on the author's farm and was due any day, we had stopped by to check and stayed to give the stock and house sitter a hand.

The ewe lamb looked fine, but she wouldn't nurse, even though the mother seemed calm and cooperative. We were also puzzled by no signs of the afterbirth, and wondered if a twin was on the way, but after four hours we gave up hope of that.

Aware of the infant's immediate need for colostrum, we tried every device we knew and had used before. The infant would suck on a finger but not a teat. The editor managed (for the first time) to strip the ewe while one of us held the lamb's mouth open, but that wasn't really very effective. We called the vet to make sure we were proceeding in the right direction. He told us the ewe could have eaten the afterbirth—though we could reach in and find out if we wanted to be sure (we didn't)—and that we should give the lamb a couple of ounces of the ewe's milk every hour or so in a baby's bottle with a nipple.

We did so, after borrowing a bottle from a neighbor. *Keep one on hand for just such emergencies.* The nipple hole had to be considerably enlarged, and although the lamb did not swallow much, it was evidently enough to prompt her to start nursing normally as soon as we put her back in the pen with her mother. Lesson: even rank amateurs can rise to the challenge of lambing!]

If a ewe will not nurse her lamb, or in the case of an orphaned lamb, you have two options: raise the lamb by yourself or try to "graft" it onto another ewe that has lost her lamb. Graft only in the case of a mother whose lamb has died. Do not attempt it on a ewe who has lost one of twins. Dip the lamb in warm water to wash away its odor, dry it, and rub it with the afterbirth of the dead lamb. Let the ewe lick it and clean it off. Skinning the dead lamb and tying the hide on the lamb may also help. Restraining the mother and applying the same methods as with a real mother who rejects her own should produce results. Keep them confined in a lambing pen for a number of days and make sure she lets it nurse and doesn't injure it.

Orphans

If a lamb must be raised without a mother, foster or otherwise, you can feed it by bottle or use a pail with a nipple on it, as you would with calves (see Table 3.5). I recommend using the pail for a number of reasons: First the lambs can feed themselves and you will not spend nearly as much time as with bottle-feeding; second, a bottle lamb becomes a cosset, which is the type of lamb that followed Mary to school one day. A cosset becomes as dependent on the milker (you!) as it would its mother and goes bananas when you leave it. The bleating will drive you crazy and when it comes time to butcher—well, could you really do it?

Table 3.5: Feeding Schedule for Orphaned Lambs

Age	Amount	Frequency
1–3 days	2–3 tbsp.	every 2 hours
4–5 days	3–4 tbsp.	5 times a day
6–7 days	$^1/_2$ cup	4 times a day
2–weeks inclusive	$^3/_4$ cup	4 times a day
3 weeks to weaning (3 months)	until full	3 times a day

It is very important to get some colostrum into the lamb as soon as possible after birth. Cow's milk and conventional milk replacers are not good for raising lambs because they are too low in fat. A minimum of 30 percent fat and 25 percent milk protein is necessary. LAMA, a Carnation-Albers milk replacer, is specially formulated for lambs. You may have to use a dropper to feed for the first few days, but get it on the pail as soon as possible.

Lambs do better on cold milk than warm because it will not sour so quickly and they will drink it for more frequent and smaller feedings. This is also easier for you. If you are lucky enough to feed it by the pail (and freezing is not a problem), let the lamb feed free choice from the pail. Be sure to change the milk daily and wash and disinfect the bucket and nipple. If the lamb won't accept cold milk, warm it at first and gradually decrease the temperature. If the lamb can be put out to pasture soon, it will help as will creep-feeding.

Pinning, or the sticking of the young lamb's tail to its anus, can be a problem for the first few days and, in extreme cases, may even cause death. Check your young lambs by picking up their tails and loosening them, cleaning off the excrement if necessary. When they are a few days old, the bowel movement becomes firmer and pinning is no longer a problem.

Castrating and Docking

Lambs, both male and female, should have their tails cut off, or docked, in the first week of their lives.

You should also castrate some rams, but not those that you will market for meat when they're 6 months old or younger or those you may keep or sell for breeding. Rams grow faster than castrated males, while the latter grow faster than the ewes.

Castrating should be done as soon as the testicles descend into the scrotum. Use of an emasculatome is preferred, particularly during fly season or where there is danger of tetanus, since there is no open wound using this method. The emasculatome, a pincer instrument, cuts off the blood supply to the testicles and they become nonfunctional and soon atrophy.

An Elastrator works similarly, stretching a rubber band around the neck of the scrotum. When the Elastrator is removed, the rubber band tightens and cuts off the blood supply and the testicles atrophy.

Finally, a sharp knife can be used. Because this creates an open wound, you should work in a clean place, to avoid infection. One person should seat the little ram on its rump, then hold a front and rear leg in each hand. The second person cuts off the lower third of the scrotum, squeezes the testicles from the scrotum, then snips them off. The wound should be sprayed with an antiseptic.

Docking is recommended. It makes mating easier for ewes, results in cleaner sheep, and lessens the chance of maggot infestations. It should be done within the first week of the lamb's life.

Any of these three tools can be used for docking. Whichever method is used, the cut should be at the third joint, about $1\frac{1}{2}$ or 2 inches from the body.

If using a knife, locate the place to cut, then push the skin toward the body so there will be extra skin to cover the stub. Twist the tail a quarter turn so the blood vessels will be cut diagonally and lessen the chance of severe bleeding. After cutting, pinch and hold the end to stop bleeding.

The Elastrator may be used. It fails to cut the tail off immediately, so there is danger of infection.

The emasculator will sever the tail of the youngest lambs, but with those a few days older, a combination of that tool and a knife may have to be used, cutting the tail outside the jaws of the tool. There's very little bleeding using this method, since the blood vessels are crushed.

If you feel uneasy about trying any of these operations, watch them done by an experienced person before you attempt them.

HEALTH

Worms are the most common affliction of sheep, and they're best controlled by pasture rotation together with a good worming program. If you change the sheep's pasture every 6 weeks, the worms' life cycle will be broken. We generally worm our sheep every 6 weeks, or every time we rotate pastures, and every 6 to 8 weeks in winter.

The worming preparations available today afford a wide range of choices, most of them of excellent quality. These medications are marketed under a variety of trade names, but there are a few major families of drugs that should be remembered at the time of purchase. These are the benzinidoles, the probenzimedazoles, the pipantel group, and Ivermectin. An ideal worm-control program incorporates a number of each of the major groups into the treatment and prevention program. One common routine is to use a member of one family of wormers perhaps 3 times in a row, then switch to another family for the next 3 times. The interval between wormings is dependent on the worm situation on the farm, but an average interval is 6 to 8 weeks.

One way to keep track of the worm load is to have a fecal exam done on a regular basis. Pick up samples of fresh pellets from around the barnyard and take them to your veterinarian for microscopic examination. This will give you a good idea of the effectiveness of a particular wormer and whether to continue using it in your operation.

In young lambs with a bloody diarrhea, coccidiosis should be considered as a possible cause. When dealing with illness, it is always possible the animal has more than one problem. Finding one possible cause of a problem doesn't mean there aren't others. So it is with coccidiosis. Often one or more additional intestinal parasites can be aggravating the illness so an already acute condition is made more serious by their presence.

Many good medications are available for the prevention and treatment of coccidiosis. In attempting to diagnose any intestinal problem where diarrhea is one of the symptoms, you should take a sample of the manure to your veterinarian for an accurate diagnosis. Livestock owners waste money by attempting to "shot-gun" treat a problem. All of the kaopectate in the world won't cure a case of diarrhea caused by coccidia even though it may give temporary relief. Further, don't wait until your patient is so far gone as to be down and out before getting a little expert advice.

Sheep are infested with a wide range of intestinal worms with formidable names and intricate life cycles. For our purposes we can say that sheep are susceptible to roundworms, tapeworms, whipworms, strongyloides, and a host of others. These worms tend to set up housekeeping at rather specific points along the intestinal tract from the stomach to the small intestine and all the way back to the large bowel.

Often limited acreage makes control measures somewhat less than ideal. Sheep are close grazers, and, as such, are particularly susceptible to picking up new infestations of parasites. Many of these worms spend part of their life cycle in or on the soil and vegetation, and sheep readily pick up new parasite loads because of their grazing habits. Most of us are locked into limited acreage and don't have the luxury of vast expanses of pasture so that we can rotate in such a way as to lessen the chance of reinfestation. Since we must survive in the situation in which we find ourselves, we must adapt to the limitations of our own farms.

Ideally, most of us would love to have a clean pasture lot for our recently wormed flock to lessen the possibility of worm reinfestation. We should have a minimum of two lots so that we can alternate on some sort of a regular basis. Anything beyond this is a bonus.

Several species of intestinal parasites cause a variety of problems in sheep. Of all of the domestic livestock, the sheep is probably the most susceptible to the adverse effects of worms and is in particular need of all possible control programs.

External Parasites

External parasites usually lead to excessive itching, a generally unthrifty appearance, and, with sucking lice, severe anemia if allowed to go unchecked.

Lice: There are two varieties of lice, biting and sucking. Both can be treated by dipping in an approved dip. There are strict federal mandates on the use of various insecticides used on farm animals. If you are unsure about a particular product, check with your veterinarian before using it. A list of approved products appears in the Appendix.

Often lice are noticed in the winter when the use of a dip is not possible. In this situation, use one of a variety of dusts that will kill these miserable pests. Animals seem to use dust bags readily if they are made available. These are porous fabric bags that can be filled with an approved dust and hung in an area where the sheep can rub against them, thus getting dust into their fleece.

Ivermectin, which can be given by injection, is particularly effective against sucking lice.

Mange: Mange is caused by a variety of small mites that often burrow in the superficial layers of the skin, causing intense irritation. The type of mange most common in this country is chorioptic mange. Most often it is seen on the legs and between the toes of infected animals. Lime sulfur dips are one effective means of controlling this parasite. Toxophene is another preparation that gives good results. Again I caution you to keep current on changes in regulations governing the use of chemicals on sheep as well as other animals whose meat or products might find their way into the human food chain.

Maggots: Maggots are one of the most disagreeable parasites to infest animals. They develop from the eggs of a fly that is attracted to moist, foul-smelling areas resulting from wounds or persistent diarrhea, especially areas where there is thick wool. These miserable creatures feed on debris and dead flesh and can have a profound effect on an animal.

In large numbers they can cause an animal to go into a shock-like condition. Unless removed, they can threaten the animal's life. Every sheep raiser should be aware of their existence and check animals that have been injured or have had surgery during fly season.

Maggots should be removed by hand, a job that no one approaches with enthusiasm, but one that must be done. Hydrogen peroxide applied full strength to the affected areas is useful in getting them up and moving. Once all the offenders have been removed, the area should be treated as an open wound, kept clean, and covered with a repellant to prevent future episodes.

A product that works well to both kill maggots and repel the flies is available from your veterinarian.

Sheep Tick: Sheep keds (sheep tick) cause skin irritation by sucking blood from the skin. When sheep are infested with keds, the wool becomes ragged and discolored. Dusting, dipping, or spraying the sheep will control these insects.

Worms: The natural resistance sheep possess against worms fluctuates somewhat; the ewe seems to have less resistance against worms just before and just after lambing. Worming should be done two weeks before and just after lambing.

Flukes: Flukes are an important sheep parasite in many parts of the United States and throughout the world. These parasites spend part of their life cycle in snails, so one way of controlling flukes is killing the snail host. If this cycle can be broken, a fluke control program can be considered to be successful.

Flukes enter the sheep's liver and, given time, they will destroy the function of this most important organ. A number of effective drugs will control flukes. Consult a veterinarian for the specifics of a control program.

Trimming Hooves

Trimming hooves once a year, usually when sheep are sheared, is important in good flock management. Set the sheep on its rump with its head resting on your left thigh. This position will normally quiet the sheep. Do not hold it in this position too long, or breathing may be hampered. Trim the excess hoof off the sides so it is nearly level with the sole. Do not cut too deeply or lameness may result. If the foot is soft, a penknife can be used for trimming; if it is harder, pruning shears should be used.

BUTCHERING

While the details of butchering sheep are beyond the scope of this book, they are one of the easiest "large" animals to butcher. After watching it done once or twice, and with the help of a good guide, most stockowners can easily do it themselves in an hour or less.

A lamb will dress out at about one-half its live weight. A grass lamb, with minor variations due to breeds, will furnish 30–40 pounds of meat when butchered in the fall. A lamb, technically, is a sheep less than a year old; a *yearling* is aged 1 year to 18 months; and *mutton* is any sheep over 18 months of age. The older sheep get (or any animal, for that matter), the stronger the meat tastes. However, while mutton may be a bit coarser, the popular belief that it is no good is a total fallacy as far as I'm concerned. Commercial mutton, maybe. But home-grown mutton or lamb is another slice of meat altogether.

4

GOATS

The dairy goat is the animal for the family that has reached its limit in paying the climbing prices for milk and other dairy products. A good doe will produce 2–6 quarts of milk for up to 200 days per year. By having 2 goats and breeding them at different times, you can have at least one of the goats milking all year and have some surplus for sale or to feed to other stock.

Those of you who are not familiar with goats and goat's milk should realize that you are in the minority—more people in the world drink goat's milk than cow's milk. Goat's milk is, in my opinion, indistinguishable from cow's milk (maybe a bit better, to quote some "unbiased" sources) and even more digestable. And don't forget other dairy products, cheese and butter, as well as leather from the hides and chevon (goat meat). All this for a fraction of the cost and space requirements of keeping a cow.

Being ruminants, goats are also ideal for our protein conversion purposes. While they have the same ability as sheep to convert low-protein feed into high-protein products, they are perhaps even more efficient feeders because they are less picky about what they consume. They are truly a browsing animal—like deer—and while they will consume grain and hay, they also delight in shrubs, weeds, saplings, bark, hardhack, and the like. They would rate number one in this book in feed efficiency, and they also rate number one in personality. If you've ever had a goat, you know what I mean; and if you're getting one, you'll find out.

BREEDS

Five recognized breeds of goats predominate in this country, and you will be considering either a purebred or some cross of one or more of these breeds.

The first three, referred to as the "Swiss Type" because of their similar conformation and origin, are the French Alpine, Saanen, and Toggenburg (Fig. 4.1).

French Alpine: Similar to the other Swiss breeds in conformation, the French Alpine can be almost any color combination. It has prominent eyes, erect ears, a straight (as opposed to dished) face, and a graceful appearance.

Saanen: Similar to the Alpine but with a dished face. Color is pure white to cream. Saanens are the highest milk producers in the group, but their milk has a relatively low fat content, a minus for those who wish to make a lot of cheese and butter, but a plus for those with dietary considerations.

Toggenburg: The smallest of the Swiss type, with the same basic conformation as the Saanen. Toggenburgs are brown to light fawn in color, with white markings in the form of stripes on the face, white ears with a dark spot in the middle, and white on the legs, near the rump and around the wattles.

The two other major breeds are American La Mancha and Nubian.

American La Mancha: A relatively new breed that has almost no external ears. La Manchas have straight faces and various color combinations. They are quite calm and good milkers.

Nubian: A popular breed with long, droopy ears and a roman nose. Nubians have a variety of colors and combinations. They are most noted for the high butterfat content of their milk, but the lowest (on the average) volume of production.

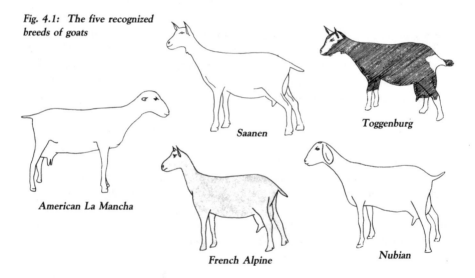

Fig. 4.1: The five recognized breeds of goats

Saanen

Toggenburg

American La Mancha

French Alpine

Nubian

PURCHASE

Within each breed you will see different types of breeding: purebred, American, recorded grade, crossbred, and unrecorded grade. A *purebred* is a goat whose parents are both registered and of the same breed. An *American* is a goat that is a result of grading up, by breeding three successive generations of a goat to purebreds of one breed. The result after three generations of breeding will be $7/8$ of that purebred and be termed an American (e.g., a grade goat bred with a purebred will produce a kid that is half purebred; the next breeding will produce a $3/4$ purebred; the final a $7/8$ purebred or American). The breed American La Mancha came about by breeding a short-eared breed with a purebred stock. A *recorded grade* is a doe (bucks of this type cannot be recorded) with one parent of unknown or mixed breed and the other a registered purebred. A *crossbreed* is the result of breeding two purebred parents of different breeds; and an *unrecorded grade* is of unknown or unrecorded parentage.

As far as breeds go, there is not enough difference between one or another to single out any as superior. If you want more byproducts such as cheese or butter, you might try a Nubian; but any breed should fit your needs. By far your chief concern should be: Is the breed of your choice popular in your locale? You may have the handsomest purebred Nubian, but if there are no others around, you'll have a problem breeding. Check with any goat clubs in your neighborhood to find out which are the most common breeds in your area.

What type of animal should you buy: purebred, cross, grade? Should you buy a kid, yearling, or older milking doe? There are no pat answers to these questions; it will depend on you. A purebred doe will not necessarily produce more than an unrecorded grade, and she will cost more. However, with recorded animals you have records of their parentage and perhaps milking records. With any unrecorded animal you are gambling. You might get some hints as to future production by conformation, but records are the best source.

Purebred goats will have more valuable offspring. You'll always have kids to sell, and, if they're purebred, the chances are you'll find a better market (both for does and breeding bucks), and they will fetch a higher price.

If you buy a 6-week-old kid you may have to wait up to a year before you get milk. A yearling will cost more but will be ready to give milk. An older doe will cost somewhere between a kid and a yearling but will have a less productive life ahead of it and is more of a gamble for all but the most experienced buyers.

Unless you have records, telling the age of goats is not infallible, but you can check the teeth the same way you do with sheep. Forget a buck to start with (more about them in the section on *Breeding*). When you look for your goat, if possible take an expert with you and do not be in a rush. Shop around. Again, going to shows or fairs and noticing top-rated animals will help you learn good points.

Body

While we look for blocky, heavy conformation in meat animals, dairy animals should have a more angular conformation: thin thighs, prominent hip bones, and a comparatively "lean" (but not sickly) appearance. Any thicknesses of the body such as a short neck or any fatty areas are indicators of poor milk producers. A good dairy goat should be rugged-looking with a broad chest and large girth and with front feet set wide apart. The back should be straight (not arched or swayed) and there should be a gradual slope from the hip bones to the tail. A wide barrel and ribcage indicate good food capacity and a potential for large litters. The ribs should be far enough apart to slip a finger in between. The skin should be smooth, thin, and supple.

Feet and Legs: The goat should stand erect and move about with ease. Check hooves for foot rot and poorly trimmed feet.

Head: The goat should have bright eyes and an alert appearance. Check the ears for any sores or scabs. Avoid a doe with any lumps or malformations of the jaw. Check the teeth to help in calculating age, and reject a doe with missing, excessively worn, or broken teeth.

Udder: Obviously, the most important part of a dairy animal. Make sure the udder is well formed and free from any injuries, scars, or extra teats (Fig. 4.2). It should have a lot of capacity and be held high enough to avoid injury from hitting objects. Look for one that is soft with no hard spots in the udder proper or teats (the udder will be firm if the doe has not yet been milked). While the udder will never be perfectly balanced, avoid one that is terribly unbalanced, and avoid a doe with overly large teats or ones that are pointed sideways. (If they look you in the eye, don't buy.) The ideal teats are the right size for milking and tilted slightly forward.

Fig. 4.2: Check the udder for a healthy formation

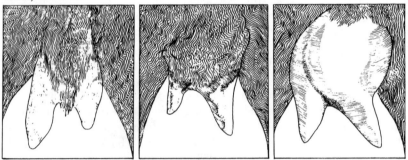

| *Teats too large* | *Unbalanced: lacks capacity* | *Well-formed* |

If you know how, milk the animal to determine capacity and temperament. See if the seller has any milking records of your doe or her ancestors. Records beyond the second generation, no matter how impressive, are of no use and can be misleading. Look for a long milking cycle and high production. Between 3,500 and 4,500 pounds per year is a good average. Records also show that the seller has an interest in his flock and strives to improve it.

HOUSING

Housing for goats is almost identical to that for sheep with a few exceptions. Goats are not so hardy as sheep and need to be more protected from drafts in the winter; and they are better at getting through fences, so more care is needed. By reading the section on Housing in the chapter on Sheep, you will get a good idea of what you'll need, with the following modifications, additions or deletions.

Pasturing

Tethering: Goats, like sheep, can be tethered but they don't take to it so well, and extra vigilance is needed. While goats are ruminants like sheep, they browse more like deer. They not only need grasses but a wide variety of feed: plants, small trees, shrubs, and hardhack. Tethering may affect their food intake adversely.

Fencing: Goats are much harder to confine than sheep and may be one of the most difficult livestock to keep where you want them. Electric fence can be utilized as with sheep, but use three wires, one each at 10, 20, and 40 inches. You'll need a strong charger and may have to train your goats to adjust to the fence. This, in my opinion, is not the best method.

The USDA recommendation of a 5-foot fence for sheep should be adequate for goats. They like to climb on the fence and if it is loose they can, in time, pull it down. Use posts at frequent intervals and get the fence as tight as possible. If this still won't contain them, you may have to resort to the absurd-sounding solution of a fence within a fence. Either use a strand of barbed wire or electric fence (better) 10 inches high and 10 inches away from the stock wire. A strand of electric wire 10 inches from the stock wire and a foot or two from the top may serve to discourage climbing. You may not need all of this, so don't let it scare you off.

Do not enclose any young trees within the fenced area or they will soon be devoured. It is even advisable that such trees be located out of the sight of your goats, lest they become too great a temptation. Goats, like sheep, are sensitive to certain poisonous plants: buttercup, bracken fern, cowslips, false hellebore, dutch-man's breeches, water hemlock, mountain laurel, sheep laurel, sneeze weed, white snakeroots, wilted or dry wild cherry leaves, dry oak leaves, rhododendron, hemlock, azalea, milkweed and locoweed, yews, foxglove, delphinium, lobelia, and lily-of-the-valley. This is not a complete list. Your state extension agent should be able to furnish you with one for other poisonous plants in your region.

Your goats should have shady areas. Gates should be tightly secured because goats are quite adept at getting them open. If you have a buck, he must be kept at least 50 feet away from does to prevent his odor from tainting the milk.

Drylots: Because fencing a large area for goats can be expensive, and because they are browsers and a field may not contain all their food needs, some people may wish to fence a small area, a drylot, and allow the goats to use it for exercise and take all their food to them. If you do this, you *must* be conscientious about furnishing them with food every day. This is also an option for those people with a very small space for grazing.

Pens

Housing is simple, very much the same as you'd provide for sheep; but be more careful to avoid drafts or dampness since goats are not so hardy as sheep. You can keep your goat (or goats) with your sheep, but be on the lookout for butting, especially if your sheep are a horned breed, since butting may damage the udder.

Allow at least 15 square feet for each goat in your pen. As with sheep, don't have any low doorsills on which a doe may injure her udder. For a larger herd, or in the case of limited space, stanchioning can be used. In this case allow at least an area 2¹/₂ feet by 5 feet for each goat.

Unless you have a very large herd, you won't want to go hog-wild in construction, so make use of what you have for your goat pens: a part of your barn, an old shed, or a garage. The best type of flooring, for sanitary reasons and ease of cleaning, is concrete. You will, however, need a lot of bedding on this to keep the animals warm in winter. Wooden flooring is fine as long as you provide enough bedding to absorb the urine before it soaks into the wood. Dirt and gravel floors, with adequate bedding, are suitable and afford excellent drainage. Any of the most common and available bedding materials can be used.

Milking Room

For goat's milk that is to be marketed commercially, many regulations govern the location, size, and fittings. Some call for an adjacent room for washing and sterilizing milking equipment, a place for cooling and storing milk, and a place for storing milking equipment. For the small operation, this is both unnecessary and prohibitively expensive. You will, however, need to take certain steps to insure the quality of your milk. As you probably know from experience with cow's milk in a refrigerator, milk collects odors from its surroundings. For this reason, you should do your milking in a room or building away from odors, and free of dust, dirt, cobwebs, and flies. You need not supply storage space for milking equipment if you store it in your house.

EQUIPMENT

You will need watering and feeding equipment, milking utensils, and perhaps a milking stand.

Hay/Grain

Goats are more picky than most animals. They won't eat off the ground and often won't eat hay that has fallen from a rack onto the ground. If you feed them with sheep, the sheep *might* clean up after the goats. A separate grain feeder is not needed because it's a good idea to feed your doe the grain ration when you're milking her. This will make her more agreeable when milking and enable you to keep an eye on her feed intake. A hay rack of the kind used with sheep is satisfactory as goats usually will not pull out and waste much hay in this type of feeder.

Waterers

Goats naturally need a lot of water to make milk, but they will not touch any water that is even slightly dirty. Buckets or troughs should be secured to prevent tipping and placed off the floor a bit so droppings will not fall in them. Keep water thawed in the winter, and if the tap water is very cold, add some warmer water to it so the goats are more likely to drink.

Milking Equipment

The following list includes milking equipment that is both necessary and optional as indicated:

> Small pail with cover (stainless steel is ideal)
> Pail and cloth for washing udder
> Milk strainer
> Storage bottles for milk
> Milk stand (optional, see below)
> Scale for weighing milk (optional)

Stainless steel pails with half-moon covers are worth the money and very easy to clean. Storage bottles that are free and completely adequate are half-gallon fruit-juice bottles or the like. While you can milk your goat on the ground while it is stanchioned, a milking stand (Fig. 4.3) can be made of wood. Note the place for the feed dish. If space is a problem, or if you have a permanent milking room, you may want to construct a foldaway milking stand (Fig. 4.4).

Fig. 4.3: Milking stand

Fig. 4.4: Fold-away milking stand

FEED

While goats are among the best browsers and will clear hardhack and other undesirable growth from pasture, don't expect them to thrive and produce milk on browse alone. They need good pasture or hay and a grain concentrate to produce large amounts of high-quality milk.

Pasture

In summer, pasture will constitute the bulk of their feed. A rule of thumb is ¹/₂ acre per goat, but this is meaningless without gauging the quality of the pasture. Use this guideline as a starting point. Watch the pasture to keep it from getting grazed too close, and watch the condition of your goats. There can be plenty of weeds and hardhack, but there must also be a high percentage of good grasses and legumes.

Goats do not like (nor will they thrive on) lush grasses alone. Put them on a lush, green suburban lawn and they will be miserable. If you do not have enough browse in your pasture, supply your goats daily with twigs, saplings, and tree prunings (especially from fruit trees). Be sure you do not feed or have in the pasture any plants that are poisonous. If you raise your goats on drylot, you must supply them with as much green and browse as they will eat. If your pasture becomes overgrazed, or in dry times when pasture wilts and dries up, or toward the end of the grazing season, you will want to supplement their pasture with good hay.

Pasture in many parts of the country is a seasonal crop, and you should plan your feeding program accordingly. Typically, pasture provides a high percentage of total feed intake in May and June. With the heat and low rainfall of July and August, the quality and quantity of feed available in most native grasses pasture decline sharply. With the return of cooler weather in September, good pasture is usually available well into November.

During the winter, hay will constitute the bulk of your goat's diet. The better the hay, such as fine legume hay, the lower your outlay for grain will be. Your goat will need less grain and can eat grain of a low protein percentage. Hay that is at least half legume and second cutting is preferred. As with sheep, second-cutting hay is worth the extra $5.00 or so a ton as there are fewer coarse stems and hence less wastage. If no legume hay is available, any good quality carbonaceous hay is suitable (see Chapter 8, "Grow Your Own . . . "). Hay should be fed free choice, but when stocking up for winter, figure 2–3 pounds per day per goat, or about a quarter-ton per goat.

Dry Does (not pregnant): Dry does can be maintained on good pasture or hay, with root crops (beets, turnips, carrots) and cabbage-family plants thrown in along with twigs from fruit trees. If the doe appears in poor condition, some grain may be added to the feed program (see Table 4.1).

Milking Does: Along with the pasture and/or hay, a milking doe will need about a quarter-ton of grain per year. This may be commercial grain (choose a grain that is palatable to your doe and of the proper protein percentage) or one of the mixtures listed below.

Any of these grain mixtures can be used interchangeably and to replace commercial grain. As with all stock, do not change the feeding regimen too often or too quickly. If you must change the ration, do so very gradually. The mineral mix mentioned earlier should always be available and, with it, plenty of fresh water.

Table 4.1: Grain Mixtures for Goats

# 4 Barley 50%	
# 1 Oats 60%	Oats 25%
Wheat bran 30%	Wheat bran 15%
Soybean or linseed oil meal 10%	Soybean oil or cottonseed oil meal 10%
# 2 Corn 35%	# 5 Oats 100%
Oats 35%	# 6 Oats 50%
Molasses 15%	Corn 50%
Protein supplement 15%	
# 3 Corn 45%	
Oats 20%	
Wheat bran 20%	
Soybean or cottonseed oil meal 15%	

For milking does, feed 1 pound of grain for each 3 pounds of milk produced (approximately 1½ quarts). One method of feeding is to keep increasing the grain slowly until peak milk production is reached. Then cut the grain back (again, slowly) as far as possible without lowering milk production.

In the summer, when goats are on pasture, you need furnish the grain only in the amounts prescribed above. In the winter, supply hay free choice and the prescribed amount of grain mixture.

Goats will do better on a number of smaller feedings per day, with a minimum of two times. The grain ration can conveniently be fed twice a day at milking time. This will give the doe something to keep her mind on and make getting her into the milking stand less of a chore. Also, if she is housed with sheep, this will ensure that she gets her full grain ration and is not bullied by the sheep.

Supplementing Commercial Feed

Any of the grain mixes previously listed will lend themselves to at-home mixing using home-grown grains. In addition, silages, root crops, and cabbages can be fed to goats at a rate of 2–3 pounds per day, which will replace one pound of hay.

MANAGEMENT

Milking

Hold it a minute! You don't just polo the goat on the milk stand (if you use one), begin milking away, and then guzzle down the product. You must first clean and disinfect all your equipment, wash the doe's udder, then milk, strain the milk, clean your utensils, and cool the milk.

Equipment Preparation: After each milking, all milking utensils must be thoroughly cleaned. First, rinse utensils in cold water and an alkaline detergent (do *not* use conventional dishwashing detergents as they will leave residues and possibly affect the taste of the milk); place the utensils in the solution and let them soak for 5 minutes. Once or twice a week use an acid detergent to remove any mineral deposits that tend to build up with the use of an alkaline detergent. After soaking the utensils, scrub them with a brush. You will see why seamless stainless steel buckets are easiest to keep clean. After scrubbing, rinse utensils in hot water, invert them, and allow to air dry.

Before your next milking, the utensil must be sanitized. Purchase a sanitizing chemical (chlorine, iodine, and ammonia compounds are most common), and follow directions that come with it. Usually, utensils should be soaked in hot water containing the sanitizing agent for 5 minutes.

General Preparation: Before we get down to the actual milking, a few more preparations should be made. The area around the udder and rear of the goat should be trimmed periodically so long or stray hairs do not fall into and contaminate the milk. Likewise, brushing your doe before milking will remove loose hairs and dirt.

If you are feeding cabbage or silage to your does, do so immediately *after* milking so that the flavor is not passed into the milk. Feeding such items immediately before milking may taint the milk.

Immediately prior to milking, wash the udder and your hands with lukewarm water containing a sanitizing solution; then dry. Use paper towels for these operations since they are less likely to be contaminated than sponges or cloths you use over and over.

If you're one of those people who have trouble tapping your fingers in sequence when you're impatient, then you may have trouble milking at first. With practice, anyone can become a good milker. Be prepared, however, to annoy your doe the first few times you try it and also to get more milk on the stand (and on yourself!) than in the bucket. Watching an experienced milker will help you learn, but if one is not available, follow these simple directions:

Be firm but gentle in your milking. Patting your doe before you start and talking to her will help you both. Grasp the top of one teat, as close to the udder as possible, between the thumb and forefinger and squeeze. This will trap the milk in the teat and prevent it from being "milked back" into the udder. Next, while pulling down on the teat gently, encircle the teat and squeeze with the middle finger, next the third finger, and finally the pinky. Done in rapid sequence, this is all there is to milking. Then do the same with the other teat, alternating teats until the udder is dry.

The first few streams of milk from each teat should be milked into a separate cup to look for "off" milk that may indicate disease.

As the milk flow lessens, pushing up on the teat and massaging the udder gently will help get the last of the milk out. After finishing milking, strip each teat by grasping the top of the teat and running the thumb and forefinger down the full length of it.

Goats should be milked twice a day (extra-heavy producers may need a third milking) as near to 12 hours apart as possible. Young does may resist milking at first, but if you are gentle and feed them grain, they should soon settle down. A full udder is not comfortable, and the doe will soon learn that you milking will give her great relief. Above all, be consistent in your milking. Try to milk and feed at the same time each day; always milk from the same side; and if at all possible, have the same person do the milking. Treat your goat kindly and act the same way every day—leave your bad moods outside the milk room. Taking your frustrations out on your doe's udder will not be an aid to good production.

Caring for the Milk: Immediately after milking, strain the milk into a storage can or other container. Milk absorbs any odors very quickly, and it must be chilled to 38°–40°F. as quickly as possible (ideally, within one hour) for best quality. Chilling cannot be accomplished fast enough in a refrigerator; submerging the milk container in cold water or ice water should do the trick. Change the water frequently since it warms, and when properly chilled, refrigerate the milk. A cool stream would do the job quickly and save the trouble of procuring ice cubes and changing the water.

To pasteurize or not to pasteurize? If you plan to sell your milk commercially, you will have to pasteurize. For home use or sale to consenting neighbors . . . well, it's up to you. Pasteurization, or the heating of the milk, kills disease-producing bacteria. Here is a simple, at-home pasteurization method (small, family-sized pasteurizers are also available from Sears and other farm and dairy supply outlets):

1. Place up to 6 quarts of milk in a glass or stainless steel kettle or flat-bottom pan. (Do not use copper, iron, or chipped enamel utensils. Copper or iron utensils may cause an off-flavor in the milk.)
2. Place a floating dairy thermometer in the milk. (Dairy thermometers can usually be obtained in hardware stores or dairy supply stores. Do not use candy thermometers, since these frequently have metal parts that may impart an off-flavor to the milk.)
3. Heat the milk rapidly, stirring constantly with a stainless steel spoon, until a temperature of 165°F. is reached.
4. Hold at 165°F. for 20 seconds, then place the kettle or pan immediately into a large pan of cold water and, with constant stirring, reduce the temperature quickly to 60°F.
5. Store the milk, well covered, in clean containers in the refrigerator at 40°F.

Goats are in their prime from 4 to 6 years of age, but with proper care some does will milk satisfactorily to the age of 10–12.

Dehorning

While some goats are born hornless, others are not. Dehorning is a good idea to prevent possible injury to the rest of your herd, other animals, or that favorite animal of all, yourself. Ideally, dehorning (or disbudding) should be done from 3 days to a week after birth; if much older, you may run into problems or have to turn the job over to a veterinarian. A goat that will have horns shows a swirl of fur on its forehead; a naturally polled kid will not show any peculiarities in the hair pattern. As a further test, check the skin on the forehead. If it is loose and moves, the kid will probably not grow horns, despite even the presence of swirled hair. Tight skin on the forehead indicates a horn, and you'd better get it out soon. Nubians often exhibit signs of horns at a later age, so be careful.

The two most common methods of disbudding are using caustic paste or a hot iron. The use of caustic paste is easier, but it takes a couple of days and, if not done carefully, may run down into the eyes and cause severe damage. I do not recommend using it, but if you do, clip the hair around the buttons and surround them with Vaseline to protect the skin. Secure the kid in a stanchion or in a disbudding box (Fig. 4.5) and carefully apply the paste.

After an hour, cover the paste with Band-Aids, then let the kid go. Be sure he can't rub on other goats, and keep him out of the rain, since water will make the paste run into his eyes. Watch closely to see if he rubs and spreads the paste.

With a hot iron, you burn the tissue around each bud for 10–15 seconds, effectively stopping the growth of the horns.

The box shown here is invaluable in restraining a kid. It is best to have disbudding demonstrated for you by an experienced person before you try it yourself.

Fig. 4.5: Disbudding box

HINGES

BREEDING

While you can get eggs from a hen without the service of a rooster, you cannot get milk from a goat without first breeding her. Does can be bred at about 7 months when they weigh 80–90 pounds. After the kidding, the kid or kids will be taken away in a few days, and the goat can be milked for about 200 days. The doe should rest for 2 months and should have been bred to freshen at the end of this 2-month period, and your cycle begins again.

Bucks

In a herd of the size most people will want—one doe or two at the most—keeping a buck is not necessary. It will cost you far more than the prevailing stud fee to keep him for the year, so there are no economic reasons for doing so.

Bucks have a characteristic odor that is, at the very best, unpleasant and which can taint the milk. If your space and olfactory senses allow, you may consider keeping a buck for convenience, particularly if you have several does. Like sheep, does do not always exhibit signs that are discernible to the human eye or nose when they are in heat. A number of false alarms and wasted trips to a breeding buck may convince you to buy your own buck. If you import your hired sire and have trouble discerning heat, you'll have to have him and his milk-tainting odor around for a while and expose him to the doe once a day to see if she'll accept him. This, too, can be a nuisance. In any event, try hiring a buck the first few times and see if it works.

If you decide to buy a buck, look for the qualities mentioned in selecting a milk goat. Be sure his testicles are fully descended. A good-grade buck can do quite well, but if you have purebred does, you should have a purebred buck of the same breed since any kids of such matings can be sold for a higher price.

Feed your buck enough to maintain his body weight, but don't let him get overly fat since this may make him a lazy breeder. Good pasture can maintain an inactive buck, as will hay and a pound of grain in the winter. Beginning 2 weeks before, and during the breeding season, allow 1–2 quarts of feed in addition to his normal ration, and provide more if heavy demands are being made of him and he looks out of condition.

While a buck is potent at about 4 months of age, don't use him for breeding until he is 7 months old. He shouldn't be used more than once every 2 weeks at that early age, but he can handle 4 or 5 does in his second year. Goats are hard to confine, and bucks that want to breed are often more difficult. Be sure to pen your buck securely away from your does when you don't want his services.

Does

Artificial insemination is becoming more widely used for goats, and you might want to check with your extension agent or local goat clubs to see if a service technician is available in your area. Like sheep, goats are seasonal breeders. They can be bred from late August to late February or as late as mid-March. They come into heat every 18–24 days (average: 21) for a period of 1–2 days during which time they'll accept the services of a buck. The gestation period is 145–155 days (average: 150). Table 4.2 will help you determine when your doe will kid.

While you should rest your goats (i.e., dry them off) 2–3 months a year, you can breed them in such a way as to never be without milk—if you have more than one doe. When they're both producing, you'll have surplus to sell or feed to other stock.

Table 4.2: Gestation for Goats (Based on average gestation period of 150 days)

When Bred In:	Will Freshen: (Breeding day less number below)* Month	Day
July	December	−3
August	January	−3
September	February	−3
October	March	−1
November	April	−1
December	May	−1
January	June	−1
February	July	0
March	August	−3
April	September	−3
May	October	−3
June	November	−3

*To determine day due to freshen take breeding day and subtract the number indicated on the table. For example, if bred July 10 doe would be due to freshen December 7; if bred November 20 she would be due April 19.

Detecting heat in does is not as easy as it may be with sows. They will exhibit much the same signs as sheep with the additional helpful hint that their milk production may fall suddenly. Coupled with the other signs, you should be able to learn to tell the heat period with some accuracy. Remember that young bucks can impregnate does at 4 months of age, so separate them from does early on or castrate those not wanted for breeding.

You can figure the approximate kidding time by consulting Table 4.2. If your doe is milking, you will want to dry her up about 2 months before she is due to kid. Do this by cutting the grain ration and not milking the doe. After a few days, the milk will be absorbed back into the body and she will be dried up. Plenty of roughages will aid her in this process. After she dries up you can gradually increase her grain to a point where she is in good shape but not too fat. First-kidders which were not milking when bred need feed for their own growth as well as that of their kids. Feed to achieve the same body state as with older does; start with 1 pound per day up to 6 weeks prior to kidding and 1^1/$_2$ pounds from that time to kidding.

Kidding

Kidding is almost identical to lambing, including the cutting back of bulky feeds a few days before kidding, to the actual birth process and possible problems. The chapter on *Sheep* goes into considerable detail in these matters and can be used as a guideline in kidding. The major exception is that multiple births are much more common in goats. Twins after the first year are very common, and triplets and quadruplets are not uncommon.

Feeding Kids: As with all stock, make sure the young kid(s) get the mother's colostrum. If you are going to let the mother rear the kids, then let nature take its course. However, since you have your goats for milk, it is unlikely you'll want to give up her milk for the 3–4 months it will take before the kids are weaned. Try not to let the kids nurse if they are to be raised artificially. First, milk some colostrum out of the mother and feed it to each kid, by bottle or eyedropper if necessary. Thereafter, milk the doe as you usually would, but don't use the milk for a few days to a week until it reaches its normal quality (feed it to the pigs or chickens in the meantime). It will take her about a month to reach the peak of her production again.

You can raise the kids on bottles (Fig. 4.6) or from pans, but the former method is preferred since less digestive upsets from the ingestion of air will result. You can use cow's milk or a milk replacer powder (as with a veal calf) but be consistent in whatever you feed. Warm the milk to 103°–105° F. and feed as much as they will consume in a 10-minute feeding 3–5 times a day. Offer grain and good second-cutting hay at the start of the second week. Water should be provided at all times. When you wean the kids from the milk depends on their intake of hay and grain and the quality of any available pasture. It is essential that all feeding equipment be sanitized before each use to prevent disease. A rack holding bottles can be utilized for self-feeding.

For your own convenience, try to sell doe kids as soon as possible after birth. Buck kids, unless you have purebred parentage, will not be in demand for breeding and should be castrated. This can be done within the first week or two, and the procedures are identical as those for sheep. Tattooing of ears is required for purebred stock, and you will have to buy or borrow a tool for it if you raise purebreds. You will want to identify non-purebred kids in a letter, but rather than buying a tattooing tool, use an ear notcher or put numbered tags in their ears. In smaller litters, or in small flocks, goats can be identified by markings or by their different colors.

Fig. 4.6: A self-feeding bottle rack

HEALTH

Mastitis, or inflammation of the udder, is perhaps the most serious affliction that can affect a milking goat. Symptoms include no milk and a peculiar gait. The udder is usually quite hard and often hot, and the doe may have an elevated temperature and be partially or completely off feed.

Mastitis is caused by a variety of bacteria that enter the udder through the sphincter at the very end of the teat. Many of these organisms are everywhere in the doe's environment, so the basis for control is strict sanitation.

Keeping the pens, yards, and pastures as clean and dry as possible will do a great deal to lower the likelihood of a doe picking up an infection when she lies down, allowing the teat ends to come in contact with the ground.

Another type of organism lives in the udder and so is spread from doe to doe at milking by the milker's hands, milking machines, or cloths used to clear the teats prior to milking. This is the contagious type of mastitis.

If a doe develops mastitis, consult your veterinarian for an appropriate treatment if you have had little experience in dealing with this disease. Often the recommendation will be to use one of the several antibiotic preparations that are available for dairy cattle. These products are infused directly into the udder through the teat sphincter.

Parasite control in goats is very similar to the programs developed for sheep. Goats are vulnerable to the same types of parasites as are sheep, although goats may be somewhat less susceptible to the debilitating effects of worm infestation.

Developing a good worming program for dairy goats requires very careful scheduling. Strict regulations limit the use of wormers in dairy animals to ensure that the medication does not show up in the milk and pose a potential human health hazard. These programs involve either the use of approved wormers or worming during the dry period. When worming is done during the dry period, the owner must be sure that the wormer he intends to use will not harm the developing kid.

BUTCHERING

The term *chevon* covers all goat meat from the newly born kid, cutting it up almost like chicken, to the old cull doe, with the meat used for sausage or stewing. Of course, you can butcher at any stage in between. The meat is quite good and tastes like lamb. Cook it the same way. Techniques of slaughter and butchering are similar to those used for sheep.

5

PIGS

Pigs have long been the victims of a poor image: they have been reputed to be unclean, unfriendly, and uncouth. They are the unwilling villains of George Orwell's anti-utopian novel, *Animal Farm*, and their name is reserved for the most despicable of people. Happily for us, none of these rumors is true. Years ago a farmer would raise two pigs, keeping one for his family and selling the other. The income from the sold hog would pay for the other and leave extra money for "payin' off the farm." Hence the more appreciative name, "mortgage lifters."

Given the chance (that is, not being confined to a pen the size of a linen closet), pigs may be the cleanest of all farm animals. They are, more importantly, one of the best converters of feed to meat and can make use of pasturing, table scraps, and garden surpluses to reduce feed costs. In only 5 or 6 months they will provide an average of 150 pounds of pork products. They also require relatively little space and care and are the smartest of farm animals. But *beware*—they are also very easy to make into pets, a decidedly negative factor if you plan to eat them. They will stand by your side for hours emitting squeals and love grunts as you scratch them. They are truly the sweethearts of the barnyard.

BREEDS

With pigs, as with most of the animals we are discussing, there is no one outstanding breed. Your choice will be based on personal preference and on what is available in your locale. Rather than trying to choose any one breed, you should

Fig. 5.1: Two major pig groupings

Fat-type

Meat-type

concentrate on finding outstanding specimens within a given breed. While certain breeds may have their own advantages and disadvantages, any well-bred pig is suitable for home consumption.

Pigs are categorized into two broad groups, lean- or meat-type and fat-type (Fig. 5.1). Years ago, when such "fatty" byproducts as salt pork and lard were more in demand, the "lard" pig was more common. In recent years, through selective breeding, pigs have lost the excess fat and most of them are the leaner meat-type. You should avoid the fat-type. Meat-type pigs are long and trim with hams and shoulders that are wider than their backs. These pigs will dress out with a lower percentage of fat. They are cheaper to raise, because less food goes into the production of fat than meat. They will still provide plenty of salt pork and lard for the average family.

The most common breeds of swine are:

American Landrace: The American version of the Danish hog that has made the Danes famous for their fine hams and bacon. They are white or pink with floppy ears and a long, lean body.

Berkshire: A medium-sized hog that is very solid and has little excess fat. It is black with white on its feet and often a bit of white on its face. Berkshires have broad faces that are slightly dished out and have medium-length snouts. While this gives them a slightly pug head, those that are extremely pug-headed should be avoided.

Chester White: A large, white hog with a medium snout and floppy ears.

Fig. 5.2: Examples of common breeds of swine

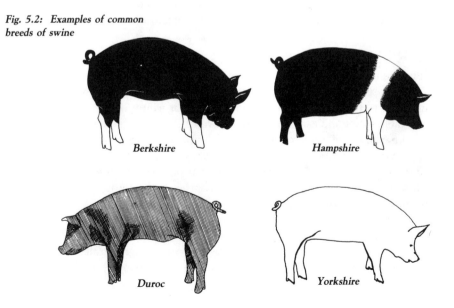

Berkshire

Hampshire

Duroc

Yorkshire

Duroc: A lean, hardy hog with large, floppy ears. It is red without a trace of any other color. The color may range from deep rust to almost tan. It tends to have relatively large litters.

Hampshire: This hog is quite popular in the Northeast and is quite easy to care for; it is a good pig for beginners. It is black with erect ears and a white band around the front of the body and forelegs and, sometimes, white on the rear feet. It is a good hog for southern climates because its darker coloration makes it less susceptible to sunburn.

Poland-China: This hog is black with white feet and splashes of white on the face and the tip of the tail. It is also particularly suited to warmer climates because of its darker color.

Spot: It resembles the Poland-China in body type but has more white. It tends to look more like a black hog with numerous white spots.

Tamworth: One of the oldest breeds of hog, the Tamworth, like the Duroc, is red, varying from light to dark, but with erect ears and a rather long, thin snout.

Yorkshire: This is a very popular breed and also easy to keep. Yorkshires are pink or white, sometimes with black spots, and have erect ears. This is the best breed, in my opinion, if you plan to go into breeding because they have large litters and are very good mothers.

It is unlikely you will be offered, or would want to buy, a purebred pig. A purebred will cost a good deal more, and unless you're planning to go into breeding, you'll be wasting your money. I have had personal experience with Hampshire-Yorkshire crosses and Yorkshire-Duroc crosses and have found them quite acceptable. In fact, pigs, like lambs, often benefit from crossbreeding by having increased vigor and growth capability (heterosis).

PURCHASE

Pigs are usually farrowed in the spring and early fall. To start, I would suggest buying a spring shoat (weaned piglet). This pig will be raised over the summer and butchered in the fall so there will be no need for winter housing. Furthermore, more table scraps and garden surplus are available in the summer, and it is easier to learn about the care for an animal during the summer. Shoats are more expensive in the spring, but it is worth the extra money to be able to raise them in the summer. It is also best to raise more than one pig at a time. If your family wants only one, see if a friend will buy another and raise it with yours. Most animals do better with companionship, and if you watch two pigs, you will notice that they never eat alone. If one gets up to eat, the other will not be far behind. In a way, they will compete for food, eat more, and tend to gain faster.

Ads in local papers and market bulletins offer young pigs for sale. Probably the best source of information is by word of mouth. A neighbor who has had a good experience raising pigs from a nearby breeder will often be a reliable source of information as to where to buy feeder pigs. These are the youngsters who have been recently weaned and are now offered for sale to anyone interested in raising one to market weight. For your first experience in raising home-grown pork, I would suggest that you go this route.

For those who fancy themselves as shrewd judges of piglets, your local auction ring will usually provide a steady supply of feeders looking for a home at which to pig out. Tracing the background on these animals is usually next to impossible. Nearby breeders often dump their excess animals here so they do not have to go through the expense of feeding them.

Buying pigs in this way puts the burden strictly on you, the buyer. Once the hammer falls, they are yours, and, except in very exceptional cases, you have no recourse if they don't turn out as you expected.

When you buy from a reputable dealer, you can ask questions about their upbringing. One might ask if any vaccinations have been given or if the piglets have been wormed.

Many breeders who sell feeders are selling a specific breed but probably not a purebred or a specific cross-breed. In most regions of the country the days of rather indiscriminate backlot breeders are gone. The pig business on the commercial level has become very specific. Crosses are developed in order that current consumer demand can be met. Today's consumer, you included, is or should be a cost-conscious individual. Pigs properly bred and fed can deliver to the dinner table a product almost as free of fat as chicken. The national Swine Producers Association has been promoting its product as "the other white meat" and has done a good job of insuring that its members retain their share of the national meat market.

Transport your little piglet to its new home in a small wooden box with some food inside it. Most pigs travel quite well (and do almost anything else well for that matter) if well fed. If you have access to a dog travel box, this will be quite satisfactory. Or you can build a simple substitute. If nothing else, they can be transported for *short* distances in a burlap sack. Do not make the mistake of thinking you can hold your cute little piglet in your lap on the way home. It would probably sour you on pigs forever.

Pigs are weaned at about 4 weeks of age and sold at 6–8 weeks of age. They should weigh 20–30 pounds at 6 weeks and 30–40 pounds at 8 weeks. Obviously, lighter pigs may be runts or less than 6 weeks of age and should be avoided.

Check the going price for pigs in your area, because it is not beneath some people to sell them to you for far above the going rate if they think you'll pay it. If you can't determine the going rate, a general formula for the price of shoats is 1.5 × weight of shoat × average price of dressed pork.

Piglets at 6 weeks of age have already been weaned from their mother and are eating dry food on their own. Iron shots, which help prevent anemia, should have been administered to the piglets at birth.

Clipping the Teeth

The needle teeth of the baby pigs should be clipped soon after they erupt to avoid potential injury to the sow's teats. Piglets are very aggressive feeders and they have no idea how rough they are while nursing. If this rough nursing takes place with sharp little teeth grasping it, the teat may become covered with minor scratches, which make the sow very apprehensive about nursing. The teeth should be cut close to the gumline with a pair of electrician's wire cutters.

There are no advantages, such as speed of growth or meat quality, in raising either a male or female pig, although some believe that males eat a lot more and consequently gain more.

Castration

All male pigs destined for the table should be castrated at about 6 weeks of age. A male that is allowed to grow to maturity without being castrated will have meat that has a strong odor and flavor that is objectionable. Once you have sniffed it, you will never again doubt its origin.

You can tell a young barrow by one or two scars found near the hindquarters and caused by the castration. If the scar is well healed it may be necessary to feel around the scrotum to be sure no testicles are present.

To determine the sex of the shoat, check the hindquarters. The female (gilt) has a small flap and the male does not. In the male, the sheath which houses the penis is quite apparent on his belly.

Look for Activity

Choose the largest and the most active pig in a litter. The belief that a pig with a curly tail is better than one with a straight tail is, in my opinion, a fallacy. Look for qualities embodied in a good meat-type hog, such as a long, lean body with large hips and shoulders.

Arrange to buy your pig early in the season and pick it up early. Go a day earlier than suggested; the seller will sell to you then and you'll get the pick of the litter. Any pig that appears listless, sickly, coughs, or shows other signs of disease should be refused. This is also true of malformed pigs or those that are humped up or whose vertebrae you can feel. A lump or bulge near the hindquarters often indicates a hernia, and although hernias sometimes regress on their own, choose another piglet to be sure. If you buy a barrow that has recently been castrated, make sure there is no inflammation or any sign of infection around the incision. If buying from a stranger, be wary of any piglet that looks different from the rest. People have been known to pass a runt on from litter to litter. That alien-looking piglet may be a year-old midget.

If you know the breeder, try to observe the pigs a few weeks before you buy so you can choose a pig the way a breeder might choose one for himself. When I want to keep a pig from one of our litters, I not only look for one with the favorable qualities listed above but try to select one that is especially alert, aggressive, and competitive for food. As with all animals, take your time when buying. Don't be rushed. Insist on choosing your own pig. You can catch a piglet by quietly stalking or cornering it and quickly and decisively grabbing a back leg and reeling it in. Ignore the shrieks and deathlike squeals—they are all just talk.

HOUSING

Pig housing can be as elaborate as you want or your pocketbook will stand, but you can get by with a small area and minimal expense. Again, start simply. Unless you live in a warm climate, buy a pig in the spring for slaughter in the fall and forget about winter housing.

Three fencing materials that can be used are wood, wire mesh, and electric fencing; and three "confinement systems" are: small dirt-floored pens, small wood- or concrete-floored pens, and large pasturing pens. Depending on what land you have available and your initiative, some systems offer substantial savings in hog raising; all, however, are suitable for home hog raising.

Fencing Materials

Pigs are one of the easiest animals to fence, but if they get out, especially if they are young, they can also be the hardest to catch. My preference for ease and economy is an electric fence. We have a small dirt-floored wood pen for our young weanlings until they are about 3 months of age when we pasture them in a large electric-fenced field. Pigs can be a problem to fence unless you are aware of the nature of the beast. If young ones escape without establishing a sense of "where the food is," you may never see them again. Older pigs rarely make an attempt to escape if adequately fed, and even if they do, they will not wander far from the site of their daily feeding. We have a friend who lives atop a mountain near us, and after butchering his feeder pigs in the fall, he allows his sow to range loose with his sheep.

Electric: This is my favorite because it is economical, portable, and quick to set up. You can buy an electric fence charger at your local farm supply store or through one of the numerous farm supply houses. There are two basic types. One is battery-operated and the other plugs into your electric circuit.

The battery-operated unit is best for fields a great distance from the house or barn where running an extension cord or other electrical line would not be cost-effective.

The plug-in type delivers a constant electrical charge to the fence and there is no worry about having to recharge the battery. When confronted with a long electrical outage, the battery-operated unit is obviously preferred.

Rather than buy metal or cedar posts, cut your own or go to the local lumber mill and buy four-foot hardwood stakes for fencing. I use half-inch steel (#4) reinforcing rods with plastic insulators attached by J bolts. This is best for portable fencing because the posts can be pulled up quite easily and the insulators can be adjusted in height to compensate for varying terrains and pigs of different ages. The metal rods can be driven into the ground with a hammer or by a simple tool constructed of two 18-inch pieces of pipe and a T coupling (Fig. 5.3). The pieces of pipe are screwed into the coupling so they form a straight handle, the hole in the coupling is placed over the rod, and you push it into the ground.

Either metal or wooden posts can be pulled up and the whole fencing moved elsewhere to make use of available feed. Set 2 strands of wire, one at 6–8 inches and the other a foot off the ground. One or two inquisitive nudges with a pig's snout and the resulting jolt will discourage the most rambunctious porker. A pig is so smart that after that first run-in it may never hit the fence again, except by accident, which will only serve to reinforce the message. After a year of electric fencing, our sow needs only one wire a foot off the ground. In using electric fencing the charger must have a good ground, and you must check regularly to be sure that weeds or rooted sod do not touch the wire and ground out the fence, thus eliminating the charge. If it does become grounded out, your pigs will soon discover this, and if they are not supplied with enough food, they will soon make an exit for greener pastures.

Using electric fencing for young pigs may be a problem because the wire must be so low that it is easily grounded. For very young pigs the bottom wire should be a bit lower than snout height and the top wire just a bit lower than the pig's height. These wire heights must be changed periodically to accommodate your rapidly growing piglets. We have had little problem fencing an entire litter of piglets in this fashion. You will find, however, that all pigs are different. Some will never go near the wire, while others will go under it all the time unless you use two strands.

Fig. 5.3: "Pipe tool" used to drive reinforcing rod into ground

— 18″ —

"T" coupling

Reinforcing rod

Wood: If you use a wood-fenced pen you will not be able to fence a large area as economically as with electric fencing. It is not portable and it takes more time to put up. However, it is far cheaper than any other form of fencing for a small pen. Don't be fancy—any scrap lumber will do. We had a collapsed chicken coop on our farm and we used the sheathing boards for our first pigpen (cost: nothing).

Drive wood posts into the ground as far apart as your boards are long and build it to a height of 4 feet. If it's much lower than that, even a 200-pound pig could scale it if so inclined. If you house young pigs, space the boards no wider than their heads (4 inches apart or so) because if a pig's head can fit through, the rest of him will have no problem following.

Pigs will root holes several feet deep so you must bury the bottom board a bit to prevent them from rooting out of the pen. Some people suggest ringing a pig's nose or running a strand of barbed wire along the ground to prevent rooting, but I think these are cruel practices. If burying the board doesn't do the trick, try burying some woven wire fencing 1 foot around the perimeter of the pen, or drive stakes along the outside of the pen about 6–12 inches apart.

Wire: I think wire mesh, unless found on your farm or inherited, is only for pigs that live on New York's Upper East Side. It is so expensive that if you invest in it you will have expensive pork chops. Pen construction is the same as for wood. Make sure the holes are small enough to contain the little fellows and keep it close to the ground or buried 6 inches and have it at least 4 feet high.

Confinement Systems

Pasturing: If given good pasture, a pig will graze just like a horse or sheep and may get up to half its food supply this way. Using the reinforcement rod posts, we make large portable pasturing pens. Shade must be provided for pigs. Either a large stand of trees or a simple lean-to or A-frame hut should be adequate. If a small stream of water is available to the pig, watering and cooling during hot weather will be simple. If not, you might have to spray your pig or sprinkle it with water during extremely hot days.

We use our pigs to rototill some rough pasture which we reseed once they have turned it over completely. If you like your pasture the way it is, make larger pens and move the pigs once the grass is eaten down or they will begin to root extensively. To guard against parasites, rotate their pastures, keeping pigs off the pasture for 2 years after grazing.

Dirt-Floored Pens: If you have a limited amount of land, hogs will do well in a smaller pen, but this will not offer you the feed economy of pasturing. Each pig should have *at least* 100 square feet, but allow 250 or more if possible. While this doesn't afford the extra food that pasturing does, the pigs will still be able to root the ground, augmenting their diet a bit and providing themselves with much-needed iron and other valuable minerals and nutrients from the soil. In a small pen the ground will be rooted bare within a short time. Toss in some fresh sod or some fresh grass clippings each day. Shade must be provided.

Wood or Concrete-Floored Pens: By pouring a concrete slab or laying a floor of 2-inch boards set a foot off the ground before fencing, you will have a pen that is almost maintenance-free and escape-proof (Fig. 5.4). A quick spray with a hose every few days and your pen will be looking (and smelling) like new. This may seem to be the lazy person's pen—but don't use this system unless you wish to invest a little extra time each day in digging sod and cutting grass. If you simply feed your pig commercial feed and some kitchen scraps each day in a pen like this, I'm afraid you'll get pork not a whole lot different from what you can buy—both in taste and in price.

In the case of wood floors, slat the boards a bit to allow wastes to fall through, but do not space them so far that a pig's foot will also slip through. As with any confinement system, shade must be provided.

Fig. 5.4: Pig pen with slatted wood floor

SLATTED WOODEN FLOOR OF 2" BOARDS SET ONE FOOT OFF THE GROUND

FEED

WATER

4'

Winter Housing

I think of "confinement housing" as winter housing because I would never use it except when I have to—in the winter.

Commercial hogs are often raised all year round in air-conditioned, humidity-controlled, heated houses with computer-controlled feedings. The method is very efficient and the product will be suitable for anyone who can't tell the difference between a fresh, vine-ripened tomato and a hothouse one. I don't believe that you can raise a healthy, good-tasting hog economically, using confinement housing year round. You can't discount fresh air, sunshine, and pasturing and rooting to get extra food and nutrients. Depending upon your locale, you may not need to confine your pigs even in the winter. Often a small, draft-free pig house set in your pen will be adequate. We have kept a pig through a Vermont winter with snow and 20 below zero temperatures in such a way by supplying her with extra hay to snuggle up in. I wouldn't suggest this in such extreme climates. We did it because we didn't have time to fix space for her in the barn. She was often stiff and chilled on some below-zero mornings, and we may have been flirting with disease.

If you expect to keep a pig indoors during the winter, plan on at least 100 square feet for a sow and her litter. The shed or barn should be well ventilated to help expel the large amounts of moisture a hog emits, but it should not be drafty. A low ceiling (preferably with hay stored above) is the best for heat-conserving reasons. And the more hogs, the warmer! Some publications suggest insulating a hog barn, but I think this is unnecessary (and expensive!). As long as your pen is free from drafts and you supply your pig with extra bedding in which to snuggle, it will be fine.

Tethering: Another form of winter confinement, much like stanchioning a cow, is tethering. It is recommended only for larger numbers of pigs, but it is worth noting. You can purchase a metal tethering collar, which is secured on the pig and fastened to the ground. The pig is able to lie down and move but is separated from its neighbor by wooden partitions. There is a food/water trough in front and a gutter for wastes behind. I don't particularly like this system because it limits the pigs, but if you have a large herd and are limited in space, I suppose there's little choice.

EQUIPMENT

You will need a feeding dish that will allow at least 1½–2 feet of feeding space per pig. It should be nontippable since pigs love to turn their dishes over, wasting food. Old iron sinks are ideal, but while they were given away a few years ago, their price now makes them something of a luxury. There is no need to spend extra money on feeding dishes (it boosts the price of your meat, remember). Use your head. I found an old coaster wagon and a flat-bottomed barbecue grill at the landfill, and after taking their frames off, I had two fine feeding troughs. I fasten them down because they are so light, but not permanently so that I can remove them for regular cleanings.

A number of self-feeders are designed specifically for pigs. They are well constructed to endure the relentless beating that pigs give them. They are expensive, and something less grand may have to do until you decide whether you want to stay in the pig business. There are some modifications of the store-bought types that you can probably fabricate for a very modest price. Try a 55-gallon drum. If you are clever with a cutting torch, you cut the drum in half the long way and make two feeding troughs. Round over sharp edges that might cut the hungry feeders, or heat them to the point where sharp edges are blunted.

Select a barrel that has not contained anything that might prove toxic to the pigs. Drums are often used to store materials that are used in the human food industry; such containers should be quite safe.

Old hot-water heaters can be used as troughs and prepared in the same way. They are usually rusty, but if you have the time and inclination to clean them, rust should not prove to be too big an obstacle.

Water is the cheapest of the essential nutrients in any animal's diet. It is particularly critical for pigs and should be presented clean, cool, and in an adequate amount. Since most pigs are fed out during the warmer months when water won't freeze, a garden hose running from the house or barn will provide the water supply. Pigs quickly learn to operate self-waterers that offer a constant supply of water from a nipple-like device. These eliminate the mess of the older style waterers. They are relatively inexpensive, have low upkeep, and are easy to install.

If you use a container, you will not want a huge one because pigs will dirty the water, which also will stagnate. A 5-gallon thick, pliable rubber pail or a 5-gallon wide-based galvanized tub is best, or any old bucket will do. Try to make use of what you have. The metal ones can be very hard to free from frozen water in the winter, and hard plastic ones will probably not last long, especially in colder weather, because pigs have a tendency, when their pails are empty, to bat them around, carry them into their houses, roll on them, and sit on them. After you have exhausted all your old pails, buy a flexible rubber one. If you have trouble with your little friends constantly spilling their water, build a wood frame around the pail or stake it to the ground.

If you pasture your pigs far away from a water source, devise a bulk-carrying storage tank that you can fill and transport to their pen, to be stored in the shade and used to refill empty buckets.

FEED

Pigs are like humans—at least in that they have only one stomach; therefore, they are not nearly so economical as multistomached animals (goats, sheep, cows) in converting low-grade protein such as forage to meat and other high-grade proteins. Then why are they in this book? For one thing, they are one of the most efficient converters of feed to meat; second, they are about the easiest animals for which to secure *supplementary* high-quality food. Because their digestive system is similar to humans, they need a balanced diet: proteins, carbohydrates, fats, mineral salts, vitamins, and water. Table scraps, garden surpluses and good pasture can supply all these requirements. With a little effort, and careful planning, it is quite possible to raise a pig using almost no commercial feed.

Commercial Feed

Consider commercial feed first because you will probably begin with it, and at times it will be all you have access to. Commercial pig feed is a "complete" feed in that it contains all the food essentials listed above. You can usually buy it in mash or pellet form, pellets often being a bit more expensive. But shop around. Local feed stores often have wide differences in prices.

Most pigs like their feed moistened so it is mushy, but experiment to find out how your pig prefers his food served. Feeding mash without wetting it may make it unpalatable and may form balls of food in the pig's stomach. In the winter, I do not wet the food, the obvious reason being that it freezes solid and robs the pig of body heat. I find that feeding a pelleted ration, dry, in the winter is the best solution.

Two pigs will grow faster and more economically than one. Never, when there are two pigs around, will only one be eating. Keep an eye out for "bullying" by one pig. If one pig is getting short-changed because of the other's "hoggishness," feed them at different locations.

From weaning age to butchering weight (about 200 pounds), you will need 600–1000 pounds of commercial feed or its equivalent (see *Table 5.1*). The most critical time in the raising of a pig is immediately after weaning. If a young pig does not get a good start, it will set a precedent for later growth. Up to 100 pounds, a pig needs feed containing at least 16 percent protein; from 100 pounds (about 3 months of age) to 150 it needs 14 percent protein; and from then on to butchering (200 pounds or so) it needs 12 percent protein.

Most commercial pig feeds are 16 percent protein so they will be fine for weanlings. A times I have raised the protein content even higher by supplementing their feed with a high-protein feed such as soybean or cottonseed meal. I have done this for 2 weeks after weaning to give our pigs an extra boost. By mixing the feed equally with cornmeal or whole corn after 3 months (at about 100 pounds) the protein content will be cut to about 14 percent. By increasing the corn content so that at 150 pounds it is two-thirds of the ration, you will have a feed that is about 12 percent protein.

Table 5.1: Alternative Feed Sources for Pigs

Weight of Pig (lb)	Approximate Age in Weeks	Amount to Feed Daily
30	7	2$^{1}/_{2}$ lb. Grade I foods.
60	12	3$^{1}/_{2}$ lb. Grade I foods.
100	17	2 lb. Grade I foods plus 3 units of Grade I and II foods in any combination.
140	21	2 lb. Grade I foods plus 4$^{1}/_{2}$ units Grade I, II, or III foods in any combination.

Grade I Foods (1 Unit = 1 lb.)

Commercial pig feed
Kitchen table scraps and swill (providing it does not contain too many potato peelings, vegetables, or too many outside leaves of greed vegetables)*
Restaurant and institution *"garbage"* (excellent if rich in gravies, meat, piecrust and bread)
Offal**
Animal afterbirth (very rich in protein)
Soybeans

Grade II Foods (1 Unit = 4 lb.)

Potatoes (boiled)
Sugar beets (raw)
Jerusalem artichokes (raw)
Belgian carrots
Corn ears on stalks
Most surplus garden vegetables (except as listed in Grade III below)
Young rape, young grass, clover, fresh clippings

Grade III Foods (1 Unit = 10–12 lb.) (May need to boil for increased palatability)

Garden waste, remains of cabbage family, pea and bean vines, vegetable peelings
Cabbage, kale
Turnips, beets, beet tops
Surplus fruit

Source: Bulletin no. 18, "How to Raise a Pig," Charlotte, VT: Garden Way Publishing Co.

*Kitchen scraps and swill weigh about 2 lb. to a quart. Solid vegetables and fruits weigh about 1 lb. to a quart. Greens, grasses, vines, etc. weigh about 14 lb. to a bushel.

**Fresh pork trimmings or fat must be boiled first (or frozen for three weeks) to kill any trichinae larvae.

A feeder pig (one that you raise for meat) should be fed as much food as it will clean up between feedings. If there is a lot of feed left over from the previous feeding, cut down to avoid waste. If none is left, increase until there is just a bit left at the time of the next feeding. I feed our pigs 3 times a day for the first 3 months and twice a day from then on. This, however, is not critical, and if 3 feedings are inconvenient, don't worry about it.

It is not economical to raise a pig beyond 200–225 pounds. Greedy ("piggy") people reason that more weight means more pork, but this is not the case. Gains over 200 pounds require more feed per pound of gain than below that weight, and gains are more fat than lean meat.

Unless you have a large scale, you'll have to estimate weight. This can be done fairly accurately and easily by using a weight tape or a weight formula. A weight tape, when used to measure a pig's girth, will read not in inches but pounds. You can purchase them in some feed stores or from farm supply catalogs.

According to Garden Way's Bulletin 18, "How to Raise a Pig," a simple and accurate formula for estimating weight is Weight = Girth × Girth × Length ÷ 400. All measurements are in inches. The girth is the measurement around the body of the pig just to the rear of the front legs. The length is from the base of the tail to a point between the ears. To make the measurements, use a tape measure or a piece of baling twine later measured against a ruler. If the weight is under 150 pounds, add 7 pounds to the total. An example: A pig measures 40 inches in girth and 43 inches in length.

$$\text{Weight} = \frac{40 \times 40 \times 43}{400}$$

$$= \frac{68{,}800}{400}$$

$$= \quad 172 \text{ pounds}$$

Supplementing Commercial Feed

While I believe you can raise delicious pork economically on commercial feed alone, a real saving for you can be realized if you supplement or eliminate the use of commercial feed by making use of unused pasture and the tons and tons of perfectly good and highly nutritious garbage that is thrown away to rot every day. It is probably true that what Americans throw away every day could feed half the world's starving people—any pig would attest to that. While my wife worked in a restaurant one year, our pig ate better than we did (shrimp, coq au vin, veal cordon bleu, etc.) on what the patrons discarded.

In today's environment we need to manage the mountains of waste that our society generates daily. Much of the waste is edible, at least edible by a pig. Many years ago every metropolitan area was ringed with swine garbage-feeding operations, often of considerable size. These operations utilized the garbage generated in the cities in a very efficient manner, converting it rather inexpensively into a highly desirable human food.

A series of situations and problems surfaced in the intervening years that largely eliminated these operations. But with appropriate technology and innovative management, it is conceivable that they could reappear. This idea should be explored regularly and it should give the backyard operator some management ideas.

Pasture: You can save up to 50 percent on feed by putting your pigs out on pasture. Not only is the pasture excellent food, but a pig allowed to range may correct any deficiencies in its diet. A pig will eat old trees, hardhack, roots, earth, and (beware) sometimes stray chickens. It will extract valuable minerals and nutrients that will enable it to make better use of its food and help keep it healthier and disease-resistant.

You cannot expect a pig to thrive solely on pasture. Pasture is not high enough in protein and does not contain everything a pig needs for growth. You will still have to supply it with commercial mash, or other foodstuffs, as will be explained later.

A young pig can be put outside as soon as the weather is sufficiently warm and it can be contained by your fencing. In warmer weather, a pig can and should be born outside. If already weaned, it will need access to high-quality, high-protein feed. An acre of good pasture will support an average of a dozen young (less than 100 pounds) pigs and 6 older pigs.

If you have poor pasture, as we did when we first moved to our farm, put your pigs to work. Confine them to a small area of pasture until they've rooted it bare. They will have also got out all the weed and quack grass roots. In extra-rough areas, where there is hardhack and small shrubs, poke a hole around the roots with a stick, and drop some corn in. Come the next day you'll have a cleared spot. After it's cleared and fertilized, move them along, remove rocks, and reseed the old pasture. In time you'll have beautiful, nutritious pasture without having to pay large sums for plowing and reseeding.

Good pasture consists of orchard grass, bluegrass, clover, and rape or a mixture. Don't put the pigs on too soon after planting and don't let them eat good pasture down too far or they'll begin rooting extensively. Good summer crops consist of Sudan grass, sorghum, and soybeans. Good late crops are rye, rye grass, and winter barley.

Garden Plowing: If because of hard rains, hail, or other catastrophes, you or a neighbor cannot harvest a crop, fence it off and set your pigs out in it. This process will make good use of an otherwise wasted crop, and cut feed costs. Even after a normal harvest a pig can be fenced in a field to glean wasted crops, stalks, and weeds. A variant of this involves the family garden. We have a strand of electric fence wire to keep out varmints and, after harvest, to keep our pigs in. They will clean out any rotted potatoes or crops left behind as well as rooting out weeds and corn stalks. They will till the garden a foot deep and fertilize it to boot.

"Garbage": This is the real harvest. Perhaps I should first explain what I mean by garbage. It is not tin cans, plastic bread wrappers, or the rest of last Christmas's mince pie found in the back of your refrigerator in July. It is *fresh* kitchen scraps, plate scrapings, and scraps from restaurants or institutions, offal from butchering, stale bread or other foodstuffs that are still edible but not relished by humans, and fresh grass clippings. I understand that commercial growers cannot feed "garbage" (raw, at least), but this is our gain. The reason it is prohibited on commercial hog farms is for disease prevention, but in a small operation it should be no problem. Disease, particularly trichinosis (more about this in the *Health* section), is caused by eating already infected pork or food contaminated with rodent droppings or the droppings from almost any fur-bearing animal. They are all trichinosis carriers. To prevent disease problems in the use of garbage, use garbage before it goes bad and keep it in covered cans or food bins away from rodents and other animals; boil questionable garbage and feed only what will be cleaned up at one feeding to prevent rodent infestation.

Boiling will never be necessary if you don't collect bad stuff and if you exercise a little restraint. That is, if you suddenly get access to half a ton of beautiful garbage, don't take it all. Take as much as you can to fill your containers and that you can use before it rots. The garbage flow will never stop unless everyone gets with it and into pigs. Our source of "garbage," a restaurant, is unpredictable (depending on its business), and we often freeze surplus in busy periods for use during lean times.

A wonderful program for making use of alternative feeds in raising pigs as outlined by The Small Pig Keeper's Council of England during World War II is given in Table 5.1. A source of milk may exist on a dairy farm. Dairy farmers withhold colostrum milk from the milk sold. Colostrum is the name given to the first milk that a cow gives after she calves. It is not suitable for human consumption and cannot legally be put in with other milk offered for sale. If farmers are not using this milk for feeding their own calves, pickling, or freezing it for future use, they may be willing to give it to you or sell it at a modest cost for you to feed your pigs.

In supplemental foods you'll have to remember that pigs are similar to humans in digestion. Feed everything in moderation. You would never eat 15 pounds of apples, nor should you feed that much to a pig. Sod fed to a confined pig is an excellent source of minerals, especially iron. For all pigs you should supply a mineralized salt block, free choice.

The Garden Way bulletin also gives these sources of diet essentials: *Protein*: meat, fish, eggs, wheat, corn, milk; *Carbohydrates*: fruit, vegetables, cereal, milk; *Fat*: meat fats, oils; *Minerals/Salt*: soil, trace mineral salt block, sod.

"Garbage" Sources: These can be as extensive as you want. You might try planting a "pig garden." Soybeans are good feed for pigs and excellent for improving garden fertility. We threw some squash seeds into a manure pile last year and got pounds and pounds of zucchini for our pig. If you throw a whole zucchini or other hard vegetable to a pig, it probably will not eat it unless near starvation. However, if you puree, mash, or boil it so that it is mushy, your pig will devour it. Also, planting extra corn will pay off, since corn is one of the best feed for finishing pork.

Most places will be happy to separate edible from nonedible refuse. As a courtesy, supply them with containers or bags and you will never lose a source. Keep an eye out for community suppers, church suppers, and the like for a source of food. Again, supply containers and offer to help. Schools, restaurants, bakeries, food stores, and neighbors are excellent sources.

Bakeries often charge for stale bread, but it is much cheaper than commercial feed and nearly as nutritious. A teacher friend of ours took a few minutes to draw a sign PIG FOOD and place it over a garbage can he placed in the school cafeteria. Some kids don't like their lunches and throw them away. More joy for your pig—and your wallet.

Sort through all garbage, discarding papers, metal, and other inedibles along with bones, coffee and tea grounds, banana peels, and rhubarb tops. Pigs will often discard what is not good for them and what they don't like anyway. You will learn your pig's tastes. Your savings will be in proportion to the time you put into collecting your supplemental food. It need not be a lot—a few minutes on the way home from work can save you many, many dollars in feed costs.

MANAGEMENT

Routines

While buying spring pigs and raising them for fall slaughter is by far the most popular routine, I'd like to say a word about the alternative, late summer to winter feeding. In colder weather a pig will only *maintain* weight on what would fatten a pig in warmer months. This means more feed is required per pound of gain. For that reason this should be considered only if you have a warm pen and an ample supply of free or inexpensive garbage. On the plus side, piglets are cheaper in the fall. A good pig house for use in the coldest months will help keep your pig warmer and cut feed consumption. Plenty of fresh water is important in the colder months to insure proper digestion.

We find such a system advantageous because we have access to plenty of fine garbage and because we eat two pigs a year but would rather not butcher them at the same time. Many pork products, namely, sausage and cured pork products, have a 4–6 month recommended freezer life, so butchering twice a year gives us plenty of fresh pork.

Handling

If you've ever tried to take your pig for a walk, you'll know why this section is here. A pig has a tendency to go everywhere but where you want it to go and often great distances in the wrong direction. For young pigs, as in picking your young shoat from the litter, corner it, grab one of the hind legs, and corral it in your arms. Holding it upside down by both hind legs often serves to quiet a shrieking piglet. As with most things, this is not guaranteed.

Larger pigs that cannot be carried or placed in burlap bags can be moved via a pig box. The box (see Fig. 5.5) measures 6 feet long, 20 inches wide, and 3 feet high and is constructed with a solid floor and slatted sides and top. The door should slide up and down in tracks formed by two ¹/₂-inch lengths of angle iron secured on each side of the door opening. Two handles extend out either end to allow handling by 4 or more people. Such a box is useful for moving a sow to be bred or taking your pig to slaughter. Building the box with skids as shown will enable you to slide the box and will make movement to a truck or from pasture to pasture easier. A few morsels of tasty food will help lure a reluctant pig into the box, but I have found after a few trips a pig will enter it of its own accord.

Fig. 5.5: A pig box

An old-time favorite pig-moving technique, and one that I've never had the slightest success with, is jamming a bucket over a pig's head and leading it backward. For some reason, when a pig is blinded in such a way it will usually quiet down and permit itself to be led backward by steering the tail.

One or more people with sticks can effectively move a pig by "herding." A few cracks of the stick on the side will correct any errors in the pig's navigation. Be prepared for a lot of legwork, since outrunning a pig is not as easy as it looks. Another method of moving a pig, for the stronger and more adventurous, is to tie a rope around a rear leg of the pig and, as it runs, pull the rear end around so it continues to run in the right direction. The pig supplies the locomotion and you, the steering.

BREEDING

I wouldn't encourage anyone to go into pig-breeding the first year in business, but it would certainly be worthwhile considering later. Stretching the mortgage lifter concept a bit, you can keep a sow and after a year or so pay her off by selling piglets, and by the second year begin generating a profit that will enable you to produce free pork for your family. In our area we never have problems getting rid of our piglets and could sell many times the number we do. Our first year we bought two gilts (females before they have produced their first litters), keeping the better one for breeding, and butchering the other. She had two litters her first year, and by her spring litter the following year, she had finished paying for herself and had begun to show a profit.

A pure bred sow is not essential but would make your piglets that much more salable if you can get a good deal on one. Yorkshires and Landraces are prolific and make good mothers, but most any breed or cross will do. A prospective sow should have a medium long body and a strong arched back. She must have strong legs, feet that are well-formed and free from injury, and she should be alert and active. Look for a large number (at least 12 or 14) teats that are prominent and well spaced.

Unless you plan on a big operation or expect to make a good income by breeding other folks' sows, don't keep a boar. It's not like sheep, where you can keep a ram for a few extra dollars a year. In our area there are not that many boars, so we have to settle for what's available. A boar selected for your sow should have good conformation and have prominent testicles of equal size. Ask about his breeding record and size of litters. Arrange to breed with the best boar in the area if the price and arrangement suit you (here's an opportunity to barter some of your extra livestock). The sow will be taken to the male. A pig box is indispensable here.

A gilt will reach puberty at 5 months of age but should weigh 200–250 pounds before she is bred. While you can breed a sow on her first heat, larger litters usually result when gilts are bred at their third or fourth heat period.

Sows that are in heat may mount other sows. The vulva becomes red and swollen. Often there is a whitish discharge. If you push on her hams from the top she may look lovey-eyed and stand there waiting. Signs vary with different pigs and they are often hard to detect, but with experience you'll be able to determine when your sow is in heat. The heat period lasts from 1–3 days (the longer period for older sows) and she'll return in heat every 16–24 days until bred.

As with sheep, *flushing*, or increasing the feed intake prior to breeding, will usually increase the number of eggs dropped and other things being equal, the size of the litter. While sows or gilt might normally be receiving 4 pounds of food per day, during flushing this should be increased to 5–7 pounds. Access to high-quality pasture is also recommended. You do not want a fat sow, but one that is both trim and gaining weight at the time of breeding.

The gestation period is, conveniently for those of us with poor memories, 3 months, 3 weeks, and 3 days (114 days). You can be reasonably certain she is pregnant if she doesn't return into heat. Otherwise it may be hard to see any other sign until about a week before, when she'll really begin to balloon and fill with milk. About three weeks before farrowing, when viewed from the side, her teats will appear enlarged and her stomach will sag a bit. When milk can be stripped from the teats, farrowing will usually occur within 24 hours. When a pig is ready to farrow, she will become very restless and begin making a nest.

If your sow has good quality, 14–16 percent protein feed, no special supplements will be needed. While a sow should be getting about 4 pounds of good feed per day before breeding and 5–7 pounds during flushing, she should be returned to 4 pounds a day for the first 3 months of pregnancy. While you want to allow for the growth of the babies, you definitely don't want the sow to become fat. At 2 weeks to 10 days before farrowing, increase the feed intake to 1 ¼–2 pounds of feed per 100 pounds of pig. The larger amount should be for gilts who are not yet mature and need feed for growing as well as supporting a litter.

On the day of farrowing some people don't feed their pigs, but I prefer to cut her feed in half. Thereafter, increase feed 1 pound per day, giving her as much as she will clean up but not exceeding 10 pounds per day. During pregnancy exercise is also important, and the sow should be fed as far from her house as possible to encourage exercise.

About a week before the sow is to farrow, you might want to prepare a farrowing crate (see *Fig.* 5.6). The purpose of the crate is to confine the babies near the mother immediately after birth and for the first few days afterward without danger of the sow sitting or lying on and crushing them. It is really a pen within a pen. The inner pen houses the sow and prevents her from sitting down too fast or plopping on her side. It is open 8–10 inches from the ground to allow the piglets to go in and out to nurse and to escape as the sow sits down. The other pen is fenced down to the floor and is used to confine the young pigs. On one side of the crate is a heat lamp.

Fig. 5.6: A farrowing crate

I have mixed feelings about farrowing crates. I think the use of them depends a lot on the sow—and you. If you have a gentle, calm sow and you plan to be around when she farrows, you may not need one. The first year we had a sow, I spent many hours building a perfectly wonderful farrowing crate. Our sow humored us, allowing me to keep her locked in it for a few days before she farrowed, but on the day of farrowing she wanted OUT. I let her out with visions of her crushing all her babies, but, lo and behold, she did quite well by herself. A lesson perhaps for all stockowners: They did quite all right before we were around and will probably continue to do so if we'll let them. I felt she would do better outside the crate if calm than inside if terribly agitated.

One advantage of the farrowing crate is that you can place orphaned or rejected piglets in with a suitable sow and she won't be able to do a great deal about rejecting them. She becomes a captive feeding station.

I think farrowing crates are best suited for larger operations where individual attention and care is simply not possible.

If you do not have a farrowing crate, you should have a pig brooder (Fig. 5.7) with a heat lamp set in a draft-free corner. Install guard rails on the remaining sides of the pen. They should extend 8-10 inches from the wall and be 8-10 inches from the floor. They permit the piglets to move about without danger of being crushed between the sow and the wall.

Fig. 5.7: A pig brooder

many hours building a perfectly wonderful farrowing crate.

HEAT LAMP

4'

8-10"

Once the sow appears to be near farrowing, she should be locked in the farrowing crate if you plan to use one. If it is her first litter, you should be around. Do not make unnecessary noise and certainly do not bring in the entire neighborhood to witness the event. A sow can become nervous with a lot of commotion and may eat her young or even hold back farrowing until she is alone. If you have acclimated your sow well, it will pay off many times over because she will probably allow you to be present and help in the event problems arise.

Once labor begins it should take about an hour or two, but exceptions abound. (Our sow has taken as long as 6 hours.) If she has one or two piglets and none for an hour or so, or if she is in labor for a long time without having any at all, leave her alone for a while. Sows are apparently shy and hold back while people are present.

If this doesn't produce results within an hour, call a vet and explain all details. The vet may prescribe an injection of oxytocin to aid delivery, or may decide that she will need a hand (literally) in farrowing. At times a piglet will become lodged in the birth canal and a little help will be needed. Wash well before and after (before to protect the sow, and after to protect yourself) and coat your hand with Vaseline or KY jelly. Bend your fingers at the knuckles so that your fingernails don't injure her and insert your hand (you'll go pretty deep). Feel around and don't be surprised at the kicking or even nipping. You should be able to turn the infant around and gently pull it out. Either direction will do as pigs are born both head and feet first.

I like to be present in colder weather to dry off the babies, help them get some colostrum, and place them under a heat lamp. A heat lamp should be used when the temperature drops below 50 degrees F. In warmer weather our pigs are born outside, and I generally leave them alone. You should remove the afterbirth, because it is not necessary for the sow to eat it and doing so might even encourage her to devour her young. If the navel cord is long and unduly hampers movement, clip to 3 inches and dip it in iodine.

A mother will usually accept her litter and the babies will nurse about every two hours. If she refuses to allow them to nurse, give her a little tap to encourage her to cooperate. If she still won't have them, you'll probably lose the litter unless you want to stay awake a few nights feeding them. If you do, goat's milk is a good substitute for sow's milk. A bottle with a nipple will not be needed because piglets can be taught to drink from a pan soon after birth.

Feed the piglets every 2 hours for the first 3 days, 6 times a day for the next 4 days, and thereafter try to keep fresh milk in front of them at all times. Allow plenty of space at the feeding bowls and be sure that the bowls are heavy or the pigs will spill the milk continually. In a day or two after birth you can offer them some grain moistened with milk—free choice. Water should be available at all times.

If the piglets did not get any colostrum, try to get some from the sow and feed a tablespoon to each piglet through an eyedropper. If you have another nursing sow, you might try getting her to adopt a few. (You can try this both in the case of an orphaned or rejected litter and with oversized litters in an effort to "even off" the litters.) In the case of a nervous sow who does accept her litter, enter her pen the first couple of days only for feeding, and do not organize tours with the neighborhood kids.

If you use a farrowing crate, you can let the sow and litter out of it in 3 days to a week, still, of course, allowing them access to the heat lamp. The light performs the double purpose of keeping the piglets warm and attracting them away from the mother after milking. It lessens the chance of any being crushed. As the weather gets warmer and the piglets get older, the light should be extinguished periodically to wean them from it. Even if the light is not turned off, they will usually wean themselves and begin sleeping outside.

Young pigs are really quite amazing, being able to scuffle and run around an hour after birth. The first day or two are the most critical in terms of being crushed, and guard rails and the removal of heavy bedding or nest material (which the piglets might bury themselves in, out of sight of the sow, and be inadvertently crushed) will go a long way to preventing losses.

Within 3–4 days you can feed the piglets some feed mixed with warm skim milk. This is especially important with large litters to take the edge off their appetites and give the sow a break. If any runts in the litter are having difficulty getting milk, we often supplement their feed with 1–3 feedings of warm skim or goat's milk mixed with a little grain. We usually slip the piglet out of the pig house, run it into our kitchen, and give it the royal treatment. We have saved many a little pig this way that I'm sure would never have made it without the little extra feed we gave it.

Water given at an early age is critical for later growth. Pigs that are not given access to water early in life will learn to do without it but may require up to 25 percent more feed for growth.

Males should be castrated at 5 days of age, because the stress is less and the mother's milk is then at its peak. There is rarely more than a trickle of blood and recovery is quick. Have on hand a new razor blade or surgical knife and a spray can of merthiolate or other antiseptic. Wash your hands well. Have someone hold the pig upside down by the rear legs with the pig's head between the holder's knees. After a few minutes of being in this position the pig will quiet down. Make an incision between the two testicles or above each one until they are exposed. Cut through the thin membrane that contains them and push the testicles with your finger until they fall forward. When each testicle has been removed from the scrotum, grasp it firmly and twist and pull on the cord until it snaps. Watch this procedure a few times and do it under supervision before attempting it yourself. So many operations appear very simple in the hands of a skilled operator but when a novice attempts it, it becomes a more complicated operation with less than desirable results.

Iron is essential in preventing anemia in young pigs. Iron is a component of hemoglobin, the substance that makes blood red. Part of its action is to assist in the oxygen-carrying capacity of blood. A sow's milk does not contain enough of it, and even if the piglets have access to sod they will still not absorb all the iron they need. Inject 1–1½ cc of iron in the fleshy part of the pig's neck at 2–3 days of age. Injecting in the neck is preferred over the rear leg to prevent lameness and staining the hams. Iron shots will reduce mortality, boost weaning weight, and enable the pigs to gain faster on less food.

At 4–6 weeks, wean the piglets from the sow. Wean the largest first and leave the smaller ones on longer to give them an added boost. You may have problems with the larger piglets accepting the late weaners, so be prepared to house them separately or else wean them all at once. Try to wean them so that they are out of sight and earshot of the mother and vice versa. This is mostly for the piglets' sake because after 4 or 5 weeks of them, the sow will be glad to get rid of them. Cut the mother's food down to normal maintenance ration at this time and supply her with plenty of roughage to aid in drying her up. In about 21 days she will come into heat and can be rebred if so desired.

HEALTH

Swine producers were one of the first groups of livestock producers to develop a sound preventative medicine program. In the 1920s the McLean County System of swine sanitation was developed to lessen the possibility of infectious diseases being passed from one generation of pigs to the next. The basic principles of the McLean County System remain as sound today as when they were first advocated. Commercial swine operations are no longer managed in this way, but certainly for backyard or small-farm operations, the principles are of considerable value.

Here is the McLean County system of swine sanitation:

1. Clean farrowing pens thoroughly and scrub them with scalding water and lye.
2. Clean the sows, particularly the udders, just before farrowing.
3. After farrowing, haul the sow and her litter to clean pasture.
4. Keep the pigs on this land until they are 4 months old.
5. Practice pasture rotation, keeping pigs off pasture for 2 successive years.

Diseases

Young pigs are particularly susceptible to roundworm (ascarids) infestations. These are worms of considerable size, and during part of their life cycles, they migrate through both the liver and the lungs. While in the lungs, they cause a condition called "thumps" in which the piglets show an exaggerated abdominal type of breathing. Pigs so affected often end up as runts.

Several good wormers are available to eliminate these parasites. Refer to the section on sheep parasites for a list of possible choices.

Several types of worms can infest pigs. Some types live in the stomach and are, quite logically, called stomach worms.

In addition to roundworms there are several other types of worms that live in the intestinal tract, and most are effectively controlled by a combination of worming on a regular schedule and good sanitation. A simple effective worming program can be as follows:

1. Treat sows and gilts 5–7 days before farrowing and again at mid-lactation (2–3 weeks).
2. Treat weaners before placing them on their fattening ration and then again 8 weeks later.
3. Treat fatteners at about 100 pounds.
4. Treat boars at six-month intervals.

Vaccination

One disease that I would strongly suggest vaccinating against is swine erysipelas or diamond skin disease. The organism that causes this disease is everywhere in the soil and can be easily prevented by a good vaccination program. The infection always seems to strike pigs just about the time they reach butchering weight. Although pigs with the disease usually respond well to antibiotic treatment, butchering must be delayed to allow the antibiotics to clear from the meat. This means extra expense for feed, medication, and perhaps labor.

BUTCHERING

A pig 5-6 months old and weighing about 200–225 pounds has reached slaughter weight. Although paying someone to butcher your pig will increase the price of your pork, you should have it done at least once and watch so you can attempt to do it in the future.

Slaughtering a pig is no small operation to be taken care of between halves of a football game. The pig must be shot, stuck (bled), dipped in hot water, and have its hair scraped off. Then it is gutted, split, and hung. After the meat is thoroughly chilled, the carcass is cut up. There are several good books dealing with this process from slaughtering to curing the meat and making sausage.

Whoever does it, however, and whatever your ultimate cost, it will be the best pork you've ever tasted.

6

VEAL

If you are tired of paying astronomical prices for veal (or have you simply not tasted it in years?), you might be surprised to learn that you can grow 100 pounds of veal in just 3 months. No other animal we are talking about, including pigs, can give you that much meat in a single-animal operation in such a short time. While you can do it economically on commercial feed alone, it will be considerably cheaper if you have access to good cow's milk.

Veal calves require a minimum of space and time and could conceivably be raised in a suburban garage without anyone knowing, if you could explain the daily supply of rich manure you have for your garden and the occasional unusual sound emanating from within (car trouble?).

This furry little ball must be butchered at three months of age. Remember that by keeping his pen clean and feeding him well, you are giving him a nicer life than he would have elsewhere. Above all, do not make a pet out of him. A veal calf you decide to keep will soon eat you out of house and home.

BREEDS

While any breed can produce acceptable veal, a Holstein is preferred. A cross of a Holstein and a meat breed is also a good choice.

Fig. 6.1: Holstein calf

PURCHASE

In purchasing a calf the biggest consideration is not necessarily *what* you buy but rather *where* you buy it. Livestock auctions are among the worst places to purchase calves. These calves are subjected to severe stress, having just been taken away from their mothers and transported under all extremes of temperature and usually crowded conditions.

Another common source of day-old calves to avoid is cattle dealers who sell them off their trucks. These calves are subjected to the same stresses that those going through the auction ring are and most are candidates for disaster.

Almost all bull calves are sold soon after birth by one avenue or another. You should be able to pick one up at a neighborhood dairy farm at the going market rate. If it is possible, have the calf stay on its home farm receiving milk for about a week.

While they are more expensive, older (preferably week-old) calves are worth the extra money. If they have made it to that age, their chances of survival are considerably better. However, you still don't know how the animal was fed or handled from birth.

Your best course is to make arrangements with a local farmer to buy one of his bull calves (heifers are fine for veal but dairy farmers usually keep heifers for replacement stock or sell them at a higher price). Farmers often have trouble getting rid of bull calves and you can usually get one cheaply. Make sure the calf gets colostrum from the mother for 3 days after birth, and see if it can possibly be fed the milk for a week before you pick it up. In most areas, farmers cannot market milk from cows until a week after they have calved. This milk would then go to waste but it can be fed to a veal calf. These animals will be healthier and more thrifty to feed.

Select a calf with a long, blocky body. Avoid those that are narrow and shallow-bodied and have long, crooked legs. You want alert calves that appear healthy and vigorous. Watch the calf for a while and try to see some of its manure. If the manure is pasty and white it probably has scours and should be avoided. Check the navel to be sure there is no infection there. Calves need not be castrated or dehorned because they will be butchered before this matters.

HOUSING

Housing for a veal calf should be no problem. The chief requirement is that the calf be kept dry and free from drafts. You can probably make use of some space you already have without having to build a pen. You can use a lambing pen, a horse's tie or box stall, or a corner of a garage or barn (Fig. 6.2). Ventilation is important in the summer so the calf does not get too hot. If the pen has windows, they can be taken out and the space covered with burlap. This will allow air to pass through but will keep the stall dark enough to discourage flies. The pen should be well bedded, but do not use any edible bedding such as old hay, straw, ground corn cobs, or corn stalks. Use wood chips, sawdust, or any other nonedible bedding. A pen must be thoroughly cleaned before a new calf is placed in it.

Perhaps on a visit to a dairy farm you have seen a calf hutch, a small shelter, often homemade, that is used to raise individual calves. Often these are constructed of weather-resistant plywood and are about 4 by 8 feet in size, solid on three sides, and have a door facing south, a slanted roof, and a very small wire enclosure in front to allow the calf limited outside access. These structures are ideal for raising calves under all but the most extreme weather conditions.

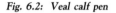

Fig. 6.2: Veal calf pen

EQUIPMENT

Equipment is minimal. You will need a dish for feed, one for water, a salt block offered free choice, and a container for feeding milk. You can use a bottle with a nipple, but in time that will not hold enough milk for a feeding, so the best bet is a "nipple pail" (sold commercially under many names, among which "Calf-a teria" is the most fitting). This will hold a large supply of milk. After each feeding, the milk pail and nipple should be thoroughly cleaned and sanitized with soap and hot water. Hang the bucket or bottle so the calf must raise its head—the more natural way to drink. Nipple feeding eliminates waste and also cuts down on gulping, which can lead to digestive disturbances and scours.

FEED

Some experts believe that feeding milk in the recommended quantities and allowing a grain mixture free choice will produce a healthier, hardier animal than a milk-only diet without being detrimental to the meat itself. The meat is still pink and fries up white and tender.

Milk replacer, which when mixed with water has the consistency of fluid milk, is sold under many brand names in feed stores. Some companies market a milk replacer made especially for a veal calf, but you can raise your calf on the common milk replacer almost as well. Follow the manufacturer's directions for mixing the powder. Usually it is mixed with a wire whisk with 115° water and served at a temperature of 105°. When the calf is 4 weeks old, consumption is greater when the milk is served at 85°. Feed twice a day as close to 12 hours apart as your schedule allows. If the package directions do not give adequate information for feeding veal calves (as opposed to nonveal calves that are being fed grain and pasture or hay), supply them with as much milk as they can finish off in 10 to 15 minutes. Do not overfeed, especially in the first few days of life (see *Health* section). The feeding of fresh milk will be covered in the section on Supplementing Commercial Feed.

A commercial or home-mixed grain comparable to 14–16 percent dairy fitting ration should be available free choice at all times if you choose to go the grain route. With their milk ration the calves' grain consumption will not be very high but will be enough to make them a little hardier. Water and a salt block should be available at all times.

If the calf eats slowly or is off its feed, check his temperature and his manure. While the manure will be looser than that of cows or calves on conventional feed, if abnormal runniness is present, refer to Table 6.1.

Table 6.1: Feeding Schedule for Calves with Runny Manure

Next two feedings: withhold all feed. Replace with electrolyte solution (see Health section).

3rd and 4th feedings: normal amount of liquid, but half of normal solids. No grain.

5th feeding: normal schedule.

If you wish to produce the ultimate product for your table, your calf will receive nothing but milk or milk byproducts. There are a variety of milk replacers on the market that are designed specifically for the veal calf.

The gourmet will insist that veal must be completely white without a reddish tinge. Your tastes may not be quite as critical, thus somewhat more tolerant of veal that doesn't meet the gourmet's criteria.

A neighboring dairy farmer may serve as a source of inexpensive feed for your vealer. Colostrum and milk from cows treated with antibiotics for a variety of reasons can be used as veal feed. In a typical situation, some combination of locally obtained and commercial products will turn out to be the feeding program that will be followed.

Exercise should be restricted as the meat produced should be tender enough to be cut with a fork.

Fresh milk is naturally much better for calves and scouring is much less common when it is used. Feed at the temperatures recommended earlier and feed twice a day as much as the calf will drink in 10–15 minutes. Stick to milk from the same farmer since switching back and forth between flocks can upset the calf's digestive system.

Calves tend to grow better at temperatures between 50 and 60 degrees, so you can set your raising schedule to the times when those temperatures are likely in your region.

You may be able to realize some monetary gain in the sale of hides to a local tanner or by selling the hide that you tan yourself. With the hair on, calf skins make fine rugs or bed coverings; without the hair, their leather is of high quality.

Handling

Handling of veal calves is not a great problem because they rarely get larger than 200–250 pounds. A halter or collar will make handling easier and they can be made to follow by coaxing with some grain.

HEALTH

The major disease or parasite problems are outlined for easier diagnosis and treatment in the Appendix. Again, strict sanitation, correct housing, and regular feeding will go a long way toward preventing disease.

Scours is the number one problem in calf rearing, but with proper management, its effects can be minimized.

Feeding a high-quality milk replacer will do much to reduce the likelihood of scours caused or aggravated by replacers of inferior or poor quality. Infants of any species benefit from the highest quality feed.

In those instances where diarrhea does develop, I have found it helpful to eliminate the milk or milk replacer for 24 hours and replace the fluid intake with one of the numerous electrolyte solutions designed specifically for that purpose that are available on the market. These preparations come in powder form and are mixed with water and fed at a rate recommended on the label or on the package insert.

The greatest hazard faced by the calf with diarrhea is dehydration. Overcoming this situation is critical if the calf is going to make it through this most serious episode. The electrolyte solutions replace the essential salts lost through diarrhea and necessary to good health as well as providing glucose for energy.

Time has proved that this treatment is usually far better than antibiotics given either by mouth or by injection.

Nothing is simple, and so it is with calf scours. There is no simple course. Several different bacteria, viruses, and parasites may cause the calf to develop the runs, and arriving at a diagnosis is seldom easy, even for the experts. Working closely with your veterinarian, you should be able to correct the condition and develop a management scheme that will do much to avoid future problems.

Isolate the infected calves and clean the pen thoroughly.

BUTCHERING

When the time comes, you will find that not having made friends with the little fellow will make the task much less of an ordeal. The optimal butchering age is 9–12 weeks. If your food supply will allow, you might try a few more weeks of feeding, but remember—if you allow the calf to get too old the meat will lose the quality of veal. At 8 weeks the average weight for a Holstein calf is 160 pounds; at 12 weeks, 215 pounds. At these respective ages Jerseys weight 102 and 141 pounds and Guernseys weigh 122 and 170 pounds. You can see why Holsteins are preferred. The calf should dress out to about 40 percent of live weight for the Holstein (slightly more for other breeds since the Holsteins are bigger-boned), or about 90 pounds of veal for a 12-week-old calf.

We cannot fully describe slaughtering and butchering here. A calf, except for boning of the legs if desired, is not a difficult animal to butcher, and watching the entire process carried out by an experienced person is the best way to learn.

That little corner of your shed that once housed your lawnmower (before you traded it in for sheep) can be put to good use in the production of your veal. Like lawnmowers, expensive veal can also be a thing of the past.

Don't forget sweetbreads!

7

BEEF CATTLE

With the ever-increasing emphasis on low cholesterol and low saturated-fatty acid intake in our diets, changes are being made in the composition of the meat products reaching our dinner tables. Not too many years ago a beef carcass that graded prime was considered the ultimate in gourmet dining. Truthfully speaking, it still is. An even fat cover and excellent marbling are still the criteria used in judging beef that will be tasty, tender, and succulent.

Nonetheless, the newer realities of our dietary knowledge have changed our buying and eating habits. Health authorities now dictate that fat consumption must be reduced to ensure our good health and to prolong our lifespans. Beef producers have developed a leaner carcass to meet these dietary recommendations. This beef animal has far less waste and can be produced more economically than the animal that grades high on the scale as prime and choice. Too, it can be raised by the small operator quite efficiently and without a major cash outlay for purchased feed.

Conventional wisdom has it that the critter that is raised at home is almost always tastier and tenderer than the animal raised in a commercial feedlot. Whether this is true is difficult to answer because comparative taste tests are rarely attempted, and the owner/raiser is almost always biased toward the product he has lovingly brought to the table.

It should be remembered that lean beef may require a different method of cooking than does meat that grades higher. The fat cover and the marbling of high-quality cuts generate the moisture that permits broiling without drying out the meat. Leaner cuts, with less fat cover, tend to dry excessively if prepared in the manner suggested for higher grades.

Generally such cuts can be cooked using moist heat in the way one would prepare a pot roast. First, sear the cut on all sides to brown, then add water to almost cover the piece, season to taste, cover the kettle, and simmer until tender.

Smaller servings are also recommended. This is important to remember when you have the carcass custom-cut and frozen at a slaughterhouse. With smaller portions and leaner cuts, it is possible to enjoy beef on a regular basis while lessening potential health problems for you and your family.

All this brings us to the question of home-raised beef. Is it feasible, economical, and worth the effort? Like so many questions, there are no easy answers. To say that the average backyard or weekend farmer can produce it more economically than beef can be purchased in the supermarket is probably misleading unless a number of unusual circumstances are at work in a household. But savings aren't the only consideration. If you get great personal satisfaction out of raising livestock and watching it grow and mature, a modest home beef operation should be considered.

Can You Eat It?

If you are thinking about attempting a home beef project, you have several things to consider. Is your family, both primary and extended, large enough to consume a beef carcass in 6 to 9 months? Assuming that your hypothetical steer dresses off at about 500 pounds, do you have enough mouths to eat that much meat within the limits of acceptable freezer life, say no more than 9 months? (Freezer life is that period during which the product is palatable. That doesn't mean that it cannot be eaten after that time. The meat simply is not of the same high quality it was when it was frozen.)

Everyone conjures up images of sizzling sirloins or tasty tenderloins grilled over a charcoal fire when contemplating this project. Remember the realities: A lot of the meat will be hamburg and stew beef, so the ingenuity and resourcefulness of the housewife or househusband will be tested. The amount of hamburg that one can expect from a carcass varies widely depending on the grade of the animal and the skill of the butcher, but the yield of hamburg will be close to 35–40 percent regardless of how the butchering is done.

A common strategy is to sell or share half a carcass with a relative or neighbor. Selling gives you the opportunity to recover some of the costs of bringing the animal to an acceptable slaughter weight.

Be sure that your freezer is large enough to hold the meat when you bring it home. Half a steer is a lot of meat. It can be most disconcerting to bring home your frozen custom cuts and find that the storage space available is inadequate. Such a dilemma is sure to tax your resourcefulness. Careful planning will help you to avoid this sort of confusion. Using the rule of thumb that a cubic foot of freezer space will hold about 35 pounds, a custom-cut frozen carcass weighing 500 pounds will take about 15 cubic feet of space.

BREEDS

It may sound strange to read that we are suggesting that you purchase a dairy breed instead of a beef breed for your project. Simply stated, the primary reason is economic. The bull calf of a dairy cow is, in the majority of cases, going to be sold as soon as he is born since he fills no real need on most dairy farms. Certainly a buyer could go to a beef operation and purchase a calf, but this approach has a couple of limitations. First, the beef operation is in the business of raising beef so if you buy a calf, the calf's mother has no real purpose. She is supposed to be the source of nutrients for the calf by reason of her supply of milk. This scheme then disrupts the normal flow of events in most beef operations.

A beef calf, if sold as a weanling, naturally commands a higher price than would a day-old dairy calf. In some of the smaller beef operations the calves, once weaned, go onto full feed in order to hasten the growth process and ultimate sale for slaughter. Many people, once they get into the flow of a beef operation, do indeed seek out a favorite beef breed to raise their sights and become cow-calf operations at some level. When this stage is reached, a careful review of the several beef breeds should be made.

There are many beef breeds in this country, each of which fills a particular niche in the industry. The two breeds that probably have the highest recognition are the Hereford and the Angus (Fig. 7.1). The Hereford is recognized by its white face and red coat; the Angus, by its compact body and all-black coat. These two breeds have provided the basic foundation stock for the cattle industry in the United States today.

Fig. 7.1: Examples of beef breed steers

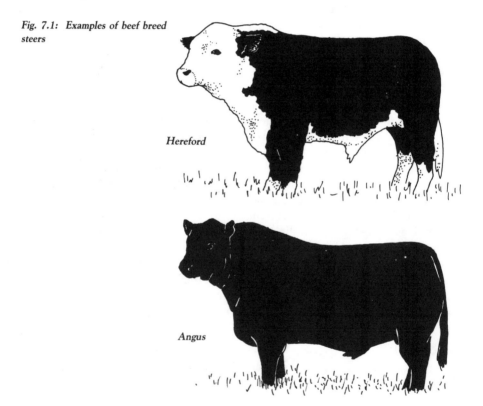

Hereford

Angus

In the post-World War II era, many so-called exotic breeds have been intro-duced into the beef industry to provide certain unique characteristics including rapid growth rate of calves and finishing a carcass at an earlier age. These breeds include the Simintal and the Charlois.

The semi-arid regions of the southwestern United States, with sparse grazing conditions and low annual rainfall, have brought back one of the breeds that formed the backbone of the industry in the early years, the Texas Longhorn. Its ability to survive and flourish under adverse conditions made it an early favorite. Today its lean carcass has prompted a revival in the interest in the breed, since that characteristic fits the needs of today's market.

Two breeds that are particularly suited to the subtropical areas of the country are Santa Gertrudis, developed on the renowned King Ranch in Texas, and the Brahma, imported from India. This latter breed is noted particularly for its heat resistance and tolerance to insects.

In the beef industry today, many of these breeds are crossed with one another to take advantage of particularly strong points of one, especially when crossed with another. A visit to a feed lot in the Midwest will reveal may cross-breeds that are specifically bred to grow to market weight in the shortest possible time.

For most of us, starting out for the first time, the basic rule is to keep the operation as simple as possible. No fancy stuff, please. In keeping with this recommendation, I suggest that you purchase a Holstein bull calf (see Chapter 6, Fig. 6.1) for your first venture into the beef business.

Fig. 7.2: Examples of dairy breed steers suitable for beef.

Brown Swiss

Holstein

I don't mean to sound prejudiced against other dairy breeds, but for beef purposes the Holstein clearly excels in terms of feed conversion. With him, you'll get the most meat for the dollar invested in feed.

The Jerseys, Guernseys, and Ayrshires do not attain the finish weight you are trying to achieve as quickly or as efficiently. Further, the so-called "yellow breeds," the Jerseys and Guernseys, possess a fat cover that is deep yellow in color. Many people find this objectionable in a hanging carcass or as a cut of beef on the dinner table.

Brown Swiss are large calves and are suitable candidates for dairy beef, but they are not as available in many areas as are the Holsteins (Fig. 7.2). Swiss tend to mature more slowly so the time required to attain suitable slaughter weight may be longer.

Why not start with a crossbreed, say a Holstein–Hereford? I say that's fine if you have access to one. Many dairy farmers still breed their first calf heifers to beef bulls, although this practice is less common today than it was even a few years ago. The reason for breeding dairy heifers to beef bulls is to try to ensure that heifers have calves of lower birth weight and consequently a smaller size, thus reducing the likelihood of calving difficulties.

PURCHASE

Dairy farmers like to get rid of their bull calves as soon as possible after birth, so most often the calves that are for sale are just a little over 1 day old. Sometimes it is possible to make arrangements with a farmer so you have the right to buy the next bull calf born on the farm. In that way it is simply a matter of the farmer calling you when the calf is born. If you choose this approach, be sure that you have made all the arrangement to take care of the calf at home. Few dairymen want to be stuck unnecessarily with one more calf to care for.

Bull calves of the beef breed are more available in some regions than others so making the decision as to which type to purchase may be based on price alone.

A word of caution to those new to the business: Beef steers, especially those of certain breeds, can have less desirable temperaments than the dairy breeds. Remember that I am taking about steers, not intact bulls. Mature dairy bulls are never, ever, to be trusted.

Don't choose any animal based on emotion. The runt, undersized and possibly malnourished, is an unlikely candidate for your first round in the beef business. If you are selecting from a group of calves of approximately the same age, use the same criteria you would in selecting a veal calf. Pick the largest, brightest, most alert calf of the group. Look for strong legs, a glossy coat, and a naturally curious attitude.

Don't rush into a hasty decision. Take your time. Visit several farms and use your natural instincts to guide you.

Where shouldn't you buy beef calves? You already know the answer from reading the previous chapter on Veal, but since education is a process of repetition, I will tell you again. There are two primary sources to avoid. Never buy a calf off the back of a cattle dealer's truck, and never buy one from the auction ring. Both of these sources are fraught with potential problems and should be avoided at all costs. Trouble is easy to find in the cattle business so there is little point in buying it.

HOUSING

Once you have decided on your purchase, you need a place to keep him. (You must forgive me for continually using the masculine gender in this discussion but it is an undeniable fact of life. Heifer calves of either dairy or beef background are destined for a higher calling in life, namely motherhood, while the males are generally consumed.)

Over the past several years calf hutches have proved to be excellent shelters for dairy calves during the earliest days of their lives. They are economical, easily constructed and maintained, and ideally suited for raising the newborn to perhaps 8 to 10 weeks of age.

A calf hutch is a three-sided box about 8 by 4 feet, with a slanted roof and preferably open to the south (Fig. 7.3). You can modify and elaborate on this basic plan. Plastic models are available that are light, durable, easily cleaned, and free of drafts (Fig. 7.4). As is so often the case, they cost more than the homemade models.

Fig. 7.3: Wooden calf hutch

The south-facing shelter usually opens into a small yard constructed of stock fencing or similar material that gives the calf a patio in which to air and frolic a bit. The calf is fed milk replacer, hay, grain, and water in this enclosure.

Less sophisticated than the enclosure but very adequate is a dog collar on the

Fig. 7.4: Plastic calf hutch

calf's neck and a light chain or a piece of baler twine of appropriate length securing the calf to his hutch.

A simple box stall in the corner of your barn, coop, or shed will also do nicely provided that its location meets a few basic needs. It should be in a bright, dry, well-ventilated, well-drained area free of drafts. If you meet these simple needs, your calf should have a suitable home.

Housing requirements for beef animals after they pass through their first several months of life are minimal. They can do well in a pasture without any housing. Having made that statement I would strongly suggest that you check the animal housing regulations in your community. Be certain that you are not violating any of them.

Most pastures have an adequate number of trees that provide shade in the summer and some protection in the winter.

When no natural protection is available during the winter, especially where you are operating in a somewhat confined, modified feedlot type of operation that involves intensive feeding for fast growth, a simple run-in shed may be necessary.

Cattle generally tolerate cold and snow very well. Cold rains make them seek shelter if it is available.

Generating body heat to offset the effects of cold, wet weather requires energy. This means that feed is consumed to create and maintain body heat. Feed used in this way is lost for use in promoting growth, so you must supply additional feed during the winter to meet this need.

FEED

Regardless of what season you start your beef, you should remember that the calf is not an efficient converter of grass into beef at an early age. In commercial cow-calf operations, the calves are on their mothers' milk until they reach weaning age. For that reason, they generally grow well early in life. Your calf will be deprived of his mother's milk so his diet is going to be somewhat artificial no matter what you do.

Stick to the spring for starting your project and increase your likelihood of success. For the first weeks of life, a calf's stomach is not the efficient converter of roughage into energy that it will become later in life. A newborn calf is essentially a monogastric animal. That means he has one functional stomach that partially converts feed into a form that can be taken up by the body and utilized for growth, energy, and replacement of tissue.

You may have heard that a cow has four stomachs. More correctly stated, a cow's stomach has four compartments in each of which certain phases of the digestive process take place.

Milk at First

The very young calf is a milk drinker. As it grows, it gains the ability to effectively utilize hay and grain for growth and maintenance. It is unlikely that any calf younger than 4 weeks effectively utilizes any feed other than milk. Somewhere after 4 weeks of age the calf's ability to use hay and grain effectively as feed increases rapidly. It is possible to remove milk or milk replacers at this time, but it is critical to continue the grain feeding to ensure that a concentrated source of nutrients is available, remembering that under natural conditions milk would still constitute a major part of the diet.

Much of what is determined to be correct in livestock management is dependent on the expertise of the owner or caretaker, and not every situation lends itself to hard and fast rules. The Bible admonishes that the eye of the master fattens the kine [cow], and so it is in this instance. Certainly if you get the impression that your animal is not growing at a satisfactory rate, some supplementation is necessary. In any instance, 2–3 pounds of a 14 percent fitting ration a day will certainly do no harm, and the worst will be that it will cost about 30 cents a day. There is a little psychological gimmick here. If your calf knows that a little treat is coming once a day, he is apt to be much more friendly and thus easier to catch when that becomes necessary.

A 14 percent fitting ration is a grain mixture that is commercially available from most feed stores that is formulated to provide 14 percent protein; stated another way, there are 14 pounds of protein in every 100-pound bag of such mixtures. This type of grain mixture is usually composed of corn, oats, soybeans, a mineral supplement, and a vitamin concentrate. As might be expected, the ingredients contained in such mixes can vary rather widely depending on the cost and availability of the various constituents and the manufacturer's own preferences. All states have regulations that ensure that the customer is getting the product as it is labeled.

In summary, we might suggest the following for a 500-pound steer for total daily intake, which is usually divided into morning and evening feedings:

Good-quality grass hay   6 pounds
Corn silage   15 pounds
14¾ fitting ration   5 pounds

For the 1,000-pound animal, the following is suggested:

Good-quality grass hay   10 pounds
Corn silage   36 pounds
14¾ fitting ration   10 pounds

Remember that these are suggestions and that there are endless combinations that you should explore with knowledgeable livestock owners in your area.

Some people put calves on grass too early. If you do, your beef project will no doubt get off to a poor start. Better to prolong the early hand-feeding program than to cause a potentially good beef animal to fail to achieve its maximum growth because of a poor start resulting from neglect or misinformation.

Even after a young animal is put on good grass pasture, continue to feed a grain supplement at some level. Remember your animal's ultimate purpose.

When he is a year old, your steer is better equipped to subsist on grass alone provided that it is of good quality and sufficient quantity.

Maximum reliance on good-quality pasture is the way to economical production (see sections on Pasture below). Without pasture, costs of feeding are going to be higher, and there is no way of moderating that factor.

Measuring Feed Needs

During the late fall, winter, and early spring, when pasture is unavailable, you must supply feed to your beefer to ensure its continued growth.

As an example, let's assume that your calf goes into the the winter weighing about 500 pounds. How much feed will it need? A rough rule of thumb is that an animal should consume about 3 percent of its body weight daily as dry matter (DM). Dry matter is the measure that animal nutritionists use in figuring out rations for the many species with which they deal.

No feed is 100 percent DM, not even the driest of hay. Most hays are about 90 percent DM. Our 500-pound animal should be eating 3 percent of its body weight daily, or about 15 pounds of DM feed. If all that is hay, an unlikely probability, then the animal should be eating 16.6 pounds of hay daily.

Here's the formula for figuring that:

$$\frac{\text{Recommended amount of 100\% DM} \times 100}{\text{DM\% of available food}} = \text{pounds of feed required.}$$

Here's how we use the figures in our example:

$$\frac{15 \times 100}{90} = 16.6 \text{ pounds of feed required.}$$

Taking this one step further, let us say that we are going to feed 5 pounds of a 14 percent fitting ration daily. If we figure 90 percent DM in the grain, then we will be supplying 5 × .90 = 4.5 pounds of DM as grain. This will lower the requirements for hay to 16.6 − 4.5 = 11.1 pounds. As the animal grows, we will make adjustments in the amount of feed being fed.

Weighing Calves

How, you may ask, am I going to know how much my critter weighs to ensure that I am feeding him an adequate amount? Ideally a scale would give you the answer. In those areas of the country where beef operations are common, portable scales are often available. These scales are an integral part of portable cattle chutes that are used to "work cattle," that is, to restrain them in order to perform necessary managerial and health procedures on large numbers of cattle. These chutes offer the restraint and flexibility necessary for range cattle that are not used to being handled on a regular basis.

If no scale is available, the next best thing is a weight tape, purchased at most feed stores. Sometimes these tapes are given away as promotional material. They are graduated in pounds. When the tape is placed around the animal just behind the front legs and snugged securely, the weight can be read directly of the tape. These tapes are accurate within about 5 percent. For most purposes that is more than adequate.

Having established an easy way to determine the weight of your animal, you can make the necessary adjustments in the feeding schedule.

While grass is usually the cheapest feed, another feed that is often available in livestock-producing areas is corn silage. It is an excellent source of energy and is readily eaten by most cattle.

To figure your animal's feed requirements, use the rule of thumb that 3 pounds of silage replaces 1 pound of hay. This being the case, you should provide 16.6 × 3 = 49.8 (make it 50) pounds of silage daily, and you should pay about one-third as much for silage as for hay, on a pound-to-pound basis.

Water

Water is the cheapest part of your feeding program. Be sure that you supply enough and that it is good quality. Perhaps you plan to use a brook to supply the animal with its water. A roaring brook in April may be reduced to a trickle in August. Monitor your water supply carefully in the extreme heat of summer, especially if you are using a pasture with which you may not be totally familiar.

In the wintertime, be certain that the drinking trough or tub is not frozen over. This requires continual surveillance on your part. Water deprivation can have serious consequences. Almost any animal can survive for rather long periods without feed, but not without water.

Several appliances are available to keep water pipes and drinking troughs from freezing. Electrical heaters and tape are the most commonly used, do not represent a big investment, and can be used for many seasons. The savings they represent in aggravation and time make them well worth buying. Going out to chop a hole in an ice-covered trough on a sub-zero morning is not everybody's idea of a good time. If you are blessed with a brook that never freezes, you are home free; but this is rarely found.

Pasture

Perhaps you are fortunate enough to have, or have access to, pasture. If you do, half or more of your animal's feed requirements can be provided. Grass, and grass alone, can produce the kind of carcass that meets all the requirements laid down by our cardiologists and nutritional advisers. Grass is, after all, the cow's natural diet. The ancestors of our highly refined and specialized cattle were strictly grazers. No other forage was available to them in the places where they originated.

People, in their usual protective and innovative way, modified the cow's diet to meet their own unique requirements whether the animals were meant for meat, milk, or draft purposes. Only in the very recent past, as evolutionary matters go, has the cow been provided with corn, soybean meal, cottonseed meal, oats, and a wide variety of other feedstuffs. When people tinker too much with the cow's diet, a wide range of digestive disorders can result. These disorders are rarely seen when the cow's diet is strictly grass and hay.

Pasture Growth

Native grass pasture tends to have growth patterns that relate to temperature and moisture levels, so the peaks and valleys of growth need to be monitored. In the northern reaches of the United States, grass tends to be plentiful in the late spring and early summer, then becomes somewhat dormant in mid-summer, to rebound in late summer and provide good grazing in the fall, and some years well into November.

Most regions of the United States have native species of grasses that are well adapted to that particular area and its seasonal rainfall and temperature fluctuations. You must become familiar with the varieties peculiar to your area in order to take advantage of the seasonal fluctuations as they occur. It is beyond the scope of this primer to identify each native species, but this in no way diminishes the importance of this understanding.

You must monitor your pasture to be certain that there is adequate feed available from mid-July to late August, and, if there isn't, to feed supplemental hay. Even a short period of water shortage coupled with excessive heat can have a most adverse effect on patterns of grass growth.

Once you have your animal gaining weight at a desirable rate on cheap feed, you don't want him drawing on his body reserves during a period of short feed.

MANAGEMENT

Avoid Stress

To ensure its best possible start in life, you should purchase your calf in the spring, ideally in April or May. Most young animals born in the wild arrive in the spring, so why break the pattern so carefully developed over the millennia?

Cattle of all ages, when subjected to changes in their environment, are stressed, regardless of the time of year these changes are attempted. When foul, wet, windy, cold weather further complicates their lives, the stress factor is multiplied, thus increasing the possibility of illness. Stress lowers any animal's resistance through a complex interaction of many factors, and this raises the animal's susceptibility to disease. Viruses and bacteria exert their harmful effects when the body's resistance is lowered through these stress factors.

Young animals generally do not have the same level of resistance to disease that their adult counterparts do, so their ability to withstand stress factors is far lower. What this all boils down to is that you should avoid moving your calf during bad weather.

There are some preventive measures that you should consider taking before moving the calf from its home farm. Ask your veterinarian what he or she would advise to help your calf make the often difficult transition from one farm to another.

We often suggest that the calf be given a vaccine prior to moving. This may lessen the likelihood of respiratory infections developing. This vaccine is administered as nose drops and is easily done in the very young calf. If you can give nose drops to a squirming three-year-old human, you can do it to a week-old calf. The vaccine is available in single-dose units and can usually be obtained from any veterinarian who has a cattle practice. Two common trade names for these products are Nasagen® and Nasamune®.

Castration

Bull calves should be castrated early in life if they are to be raised for beef. Left intact, they will begin to feel their oats at about a year and a half of age. While they're seldom mean at that age, they can prove to be more than a handful. Their size makes them somewhat less than suitable playmates. Puppydogs they are not.

Livestock scientists have conducted studies as to whether castration of bulls being raised for beef improves their ability to convert feed into meat more efficiently. My conclusion is that the jury is still out on this decision, but for backyard or hobby beef, these animals should be castrated to ensure that they are as placid as possible. Having an animal castrated ensures that his concentration will be on the task of eating as much as possible to produce the greatest possible weight gain in the shortest possible time.

There are two methods of accomplishing this: one is surgical, the other is nonsurgical.

In the surgical approach, the testicles are removed. This results in little or no setback for the animal if performed when he is quite young. The scrotum is cut to reveal the testicles. Each one is grasped and cut off with an instrument especially designed for this task, or the testicle is pulled from the scrotum until the cord is exposed. Then the cord can be twisted over and over, at the same time pulled until the testicle breaks free.

Obviously no animal in his right mind is going to stand quietly while this little exercise is taking place. It is necessary to restrain the animal while this is going on. Clever operators, usually experienced livestock owners or veterinarians, can get the job done on an animal up to 2 months old by having an assistant hold the animal's tail up in the air while they do the necessary manipulations. This little trick is not for the inexperienced or the faint of heart.

Typically the calf is cast in order to perform the operation. This means it is hog-tied (the four legs are tied together) to prevent it from either running away or kicking the person doing the surgery.

This method is best learned from a person with experience.

Another method is to restrain the animal, then cut off the bottom third of the scrotum, break down the connective tissue between the scrotum and the testicle, and expose as much cord as possible. Then, using an instrument called an emasculator, the cord is severed. The emasculator has a special crimping edge that crushes the cut edge of the cord, thus sealing the vessels that supply blood to the testicle.

The non-surgical technique is to use elastic bands called Elastrators that are placed between the testicles and the body wall. Especially made elastics are available for this and can be purchased at most livestock supply stores. A special applicator is also available which places the elastic in the desired location. Over a period of several days to a few weeks, these bands restrict the flood of blood to the testicles, causing them to atrophy.

This method is easy for beginners. One word of caution: Make certain that both testicles are in the bottom of the scrotum. Not infrequently we see situations where one testicle is above the elastic tucked up close to the body wall and the other is in its intended location.

Whichever method you decide to use, you should call your veterinarian before trying it.

HEALTH

The greatest hazard facing the newborn calf is scours, still the number one killer of young calves. If you can avoid this problem by closely following the guidelines in the chapter on veal calves, you have probably overcome the greatest single obstacle that you will encounter from a disease standpoint.

Most scours in very young calves are caused by a specific bacterium named *Escherichia coli*, an organism that can be found almost everywhere in nature and most especially in areas where cattle are housed.

A standard practice among dairymen is to dip the navel of every newborn calf in

7 percent tincture of iodine to help in preventing this organism from gaining entrance into the body by way of the umbilical vein. Although this practice has been touted for years as a tremendous aid in preventing this disease and almost all cattle raisers are aware of its importance, there are still thousands of calves born every year who suffer for lack of having this very effective, time-proved technique done. When purchasing a calf, find out if this very simple procedure has been done.

Colostrum

You should also make sure that the calf you are purchasing has received the antibody-rich colostrum from its mother as soon as possible after birth. This ensures that the animal is protected as well as possible against the large number of potential disease-causing bacteria and viruses that are always lurking in any environment. Failure of the calf to receive this critical meal often lays the groundwork for a very rocky start.

One scours-causing organism that is neither a virus nor a bacterium is called a coccidia, which logically enough gives rise to a disease called coccidiosis. This one-celled organism invades the intestinal tract and causes a diarrhea. It can be diagnosed only by a microscopic examination of the manure. Any veterinarian can do the examination for you and prescribe a course of treatment.

Periodic fecal examinations over the life of your beefer will do much to ensure that he is never heavily parasitized. Worms, allowed to go unchecked, will do much to reduce the efficiency of feed conversion in any animal and prevent the proper rate of gain. Worming cattle today doesn't have to be an ordeal. There are a number of ways the job can be done. Perhaps the most convenient of these is the pour-on form of medication. The owner pours a measured amount of liquid, based on body weight, over the back of the animal. It is absorbed and eventually reaches the intestinal tract where much of the worm population congregates. There it goes about the business of killing the worms that are present.

Respiratory Disease

A disease that always seems to be present wherever cattle are kept is the bovine respiratory disease complex. This complex is a group of viruses acting singly or in combination together with a bacterium that causes respiratory diseases in cattle that may range in severity from mild to acute, with death not uncommon if appropriate treatment is not given. As mentioned earlier, there are a number of very effective vaccines available that can control this complex. A check with your veterinarian will assist you in determining the best one suited to your particular area.

Respiratory diseases are most often seen in the late fall, when weather conditions become very changeable. They are also seen when the animals are subjected to any undue stress such as shipping, thus giving rise to the term long associated with this syndrome, shipping fever.

To restate a point, I would strongly suggest not purchasing your animal when weather conditions are not the best. I would also suggest (and insist if you were my client) that all purchased animals be vaccinated well in advance of being moved. Don't think that you can vaccinate them one day and think that they are protected for movement the next day. Most vaccines take about 2 weeks to reach their protective maximum, so plan accordingly.

If you are so unfortunate as to purchase your animal and then have it come down with a respiratory condition, prompt recognition of this and immediate initiation of treatment is absolutely critical to a successful outcome. Left untreated, the animal faces unfavorable consequences, with the worst possible outcome being obvious and the less obvious being left with chronically damaged lungs that will prevent the animal from ever realizing its full potential. These animals are generally undersized, unthrifty in appearance, and have a chronic cough. When you subject them to any undue exertion, they tend to fade rapidly and may collapse from the effort.

Treatment consists of antibiotics and good nursing care. The latter includes adequate shelter and access to the best quality feed. There should be a very minimum of handling. This is difficult, for how can you medicate animals and at the same time avoid stirring them up more than is absolutely necessary? That I leave to your ingenuity and resources.

I mentioned that in many instances respiratory disease is the result of one or more viruses acting together with a bacterium. At this time there are no commercially available medications in cattle practice that are effective against viruses so we use antibiotics against the bacteria. There are several antibiotics that can be used, including the old standby penicillin. I strongly suggest again that you contact your veterinarian to enlist his or her advice in combatting this potentially very serious disease.

Generally speaking, the isolated backyard beef animal is not at high risk to contracting a respiratory condition, for it is usually contact with large numbers of other animals from a wide variety of different environments that gives rise to the disease. In no way do I mean this to suggest that I do not want your animal properly vaccinated, because I do.

Pink Eye

This is an eye condition seen in cattle that occurs during the fly season. The bacterium causing this disease is generally spread from an infected animal to a healthy one by flies.

Early in the course of the disease you will notice the animal exhibits a sensitivity to light by squinting, plus an excessive flow of tears with visible staining of the hair on the nose just below the inner corner of the eye. The membranes on the edges of the eye socket become severly inflamed, and the surface of the eyeball may become cloudy or visibly whitened depending on the severity of the condition.

The most severe threat is that the eyeball may rupture as a result of an ulcer on the surface penetrating through to the anterior chamber of the eye with the release of its fluid content. When this occurs, the globe collapses partially or completely, resulting in blindness.

Often animals on pasture are not observed as closely as those that are confined. Further, because the early stages of this condition are rather insidious, it is possible for even the most astute, seasoned observer to miss seeing this condition. Thus it is important for the owner to check animals on a regular basis, however cursory that observation may be.

This disease lends itself to treatment using a variety of antibiotics including penicillin, nitrofurazone, and gentamycin. Your veterinarian should tell you what gives the best results in your area. The types of preparations used include drops, ointments, and injections of the subconjunctiva. The last procedure should be left to a veterinarian.

Foot Rot

Foot rot or infectious pododermatitis is a disease of the hoof that can be seen at any season of the year, although it is generally thought to be more prevalent during wet seasons. It is my experience that it can happen at any time.

At least 3 different bacteria cause this condition, working together to cause the final result. In the typical case, the condition comes up quickly with severe lameness and a hot, swollen hoof the most commonly observed symptoms. As you might expect, the animals are reluctant to move and do so with great difficulty.

When you find an animal showing these symptoms, examine the hoof to be certain that you are not overlooking something as obvious as a nail in the hoof or a stone or pebble lodged between the claws. With a case of true foot rot the tissue between the claws is swollen and, as the condition advances, that tissue will become necrotic (rotten) so that it sloughs out of that space, leaving a rather deep fissure.

There are several ways to treat this condition. The simplest is using penicillin for 3–5 days at a rate suitable for the size of the animal being treated. Check with your veterinarian for appropriate dosages and additional suggestions.

I like to wrap the feet of affected animals, usually using a salve such as iodex or ichthamol over the affected area. 2- or 3-inch wrapper gauze followed by water-proof tape works well. Use care in not wrapping the hoof too tightly. One way to avoid this is to use a pad of cotton between the toes to act as a spacer.

Hardware Disease

Hardware disease probably has been affecting cattle since the Iron Age when people first began to discard their scrap metal in places where cattle grazed. Cattle, being possessed of great curiosity and very poor judgment, are predisposed to lap at any and all objects that fascinate their minds. In doing so, they are apt to swallow a wide range of materials, often to their own detriment.

The condition gets its name from the wide variety of metallic objects that can be retrieved from the second compartment or reticulum of the cow's stomach. When surgery is performed, it is not uncommon to recover nails, wire, staples, and sundry other objects that are potentially harmful.

The clear message is that every effort should be made to explore for such materials in every paddock, pasture, or lot where cattle are kept. When doing jobs such as roofing or fence-mending, take great care to ensure that nothing is left lying around that cattle might discover and swallow.

What typically happens is that the animal swallows a nail, which falls into the reticulum. During the normal course of things, this muscular compartment is churning away as it exerts its mixing action on the contents, which are generally the consistency of a thick gruel. The inner surface of the reticulum resembles small squares when viewed from the surface. It is this part of the stomach that is the tripe often used as food.

When a sharp metallic object is caught up in the contractions, it may penetrate the inner and outer linings of the reticulum. This can quickly lead to a regional peritonitis. Stomach contents seep through the hole, further complicating the situation. Carried to its extreme, a generalized peritonitis can result. Peritonitis is an inflamation of the lining of the abdomen.

This series of events leads the animal to stop eating, since the infection causes a rise in temperature and a cessation of stomach movements which is partially due to the pain associated with the puncture.

Typically animals so affected run a low temperature, cease eating, assume a rather characteristic "humped up" attitude, and are reluctant to move.

In the worst case, the penetrating object can go forward through the diaphragm and penetrate the heart—not a good situation.

Treatment often is surgical. An incision is made in the rumen and the surgeon then reaches forward into the reticulum and retrieves the nail or other penetrating object.

A more conservative, less expensive approach is to put the animal on antibiotics for several days to attempt to overcome the infection. Additionally, a magnet is given with the hope that it will find its way to the object and prevent it from penetrating any further. At the same time a laxative/rumen stimulant is often given in an effort to get the rumen contracting again and to move out the food material that has stopped moving as a result of the peritonitis. Finally, the animal is kept as quiet as possible for several days, thus avoiding unnecessary motion and possible further penetration of the object.

Magnets are also given before they are needed to prevent objects that may be swallowed from doing their dirty work in the first place. This is a practice you should discuss with your veterinarian.

BUTCHERING

As your animal approaches 1,100 pounds, you should make arrangements to have it slaughtered. There are several ways to proceed with this final step of your beef operation.

The more adventurous of you may want to tackle the job yourself. I would caution against this unless you have a considerable reservoir of experience to draw on. Killing and dressing a steer in your backyard is a job that demands a commitment of talent and time usually beyond that of most part-time operators. Better to contract this job out to someone with the proper facilities and experience.

Most rural communities have one or more plants that do custom slaughtering, cutting, and freezing. Check with customers to determine their satisfaction with any particular slaughterhouse. By asking around the community you will get a reasonable idea of who is good and who isn't.

Getting your animal to the slaughterhouse often can be arranged with your local cattle dealer, who probably has the proper equipment to get the loading done quickly and efficiently. There is no need to make the last chapter in your beef project a rodeo. The charge for this taxi service will no doubt be nominal, but whatever it is, it is probably better than something you could rig up yourself. Most cattle dealers are making daily runs to the abattoirs in your area, so it is unlikely that your request is going to put one out of the way. Check out this method before you try loading your prize into the back of your pickup.

After slaughter, beef is typically hung in the cooler for a period of time to allow it to cure. To attain the proper level of tenderness I would suggest a minimum of 10 days for curing.

After curing, the carcass is cut up according to your specifications. Knowing the appetite and likes of your family should allow you to suggest specific sizes of roasts, steaks, and packages of hamburg and stew meat. A good butcher will accommodate your requests exactly.

As soon as they are wrapped and labeled, the cuts are flash-frozen. The freezing completed, the plant operator will call you to pick up your meat.

I have taken you from the purchase of a calf to turning to your freezer with customized cuts for your table. Good luck is not necessary in this operation. You must only follow the few basic principles I've discussed.

8

~~~ **GROW YOUR OWN . . .** ~~~

The person assigned the pleasant task of revising this book has a different outlook on the raising of field crops by the part-time farmer than did the original author.

There are, in any project, economics of time, money, and land that suggest there are practical approaches to any situation that need careful consideration before going ahead into the great unknown.

Most of us involved in the production of livestock on a limited scale do so with definitive objectives in mind, but few of us have the luxury of unlimited time to devote to our projects. Once you choose to embark on a small-scale cropping program, you must commit yourself to a great deal of hard work. This can be enjoyable, but the hard facts are that we might better channel our energies in other directions.

Given that most of us small-time livestock enthusiasts are frustrated farmers, there are realities we should face. On a limited number of acres, the amount of feed that can be produced will be considerably less than your livestock will consume. To further preserve the continuity and flavor of the original book, I have chosen to leave the author's comments and recommendations relating to the production of certain basic field crops in, at the same time cautioning the reader to realize the limitations these cropping programs have.

What we are concerned with are *field crops* and *forage crops*. Field crops are those plants that are primarily grown for their seeds: corn, wheat, oats, soybeans, and even sunflower seeds. Forage crops are the plants or parts of plants that are used for feed before maturing or developing seeds. These forages are fed as pasture (the easiest way—the animal does the harvesting itself and there are no curing or storage problems), as hay or as silage. The forages are broken down into legumes and grasses.

FIELD CROPS

Corn

Of the five types of corn, flint and dent corn are those that should be considered for feed purposes. Dent corn is the most widely grown feed corn but flint corn, because it grows more quickly, is used in regions with shorter growing seasons.

Plant the corn in rows 40 inches apart, with the seeds planted 5 to a hill and the hills 40 inches apart. Do not pick feed corn as you would your garden variety sweet corn. Wait until after a frost, when the husks are dry and the kernels firm. For small-scale operations, hand picking is the rule. A husking tool will make the job faster. The stalks may be saved as they make good bedding for other livestock.

We feed the whole ear (sometimes the entire plant) to the pigs. For other animals you must husk the corn and then remove the kernels from the ear with a corn sheller. A hand sheller can be purchased inexpensively from farm supply catalogs or from local feed stores. Whole corn can be fed to stock as indicated or it can be cracked or made into mash and mixed with a particular feed ration. Corn, plant and all, can also be used for silage. *Yield:* 75–100 bushels per acre.

Oats

Plant oats 2 inches deep and 4 inches apart in rows that are also 4 inches apart. Oats should be harvested in the dough stage when they are soft but not mushy. The grain heads should be full but not so dry that they'll fall out. Cut with a scythe, preferably with a cradle attachment to catch and help pile the oats, and rake into windrows and let dry for a day or two. Then tie them in bundles and let them cure in the field for as long as it takes them to be thoroughly dry (2 days to 2 weeks, depending upon the weather). Then the grain must be threshed (the grain separated from the plant) and winnowed (the chaff separated from the grain).

To thresh, place the grain 6–12 inches deep on a clean floor (or on a tarp on a not-so-clean floor) and beat the plants with a flail until the grain is separated from the plants. (A flail is a wooden rod about 5 feet long with a joint made of chain about $2/3$ of the way down.) After flailing, take a pitchfork and scoop out the straw (good for bedding), being sure to shake it well and release any oats left in it. Winnowing is not necessary for oats that are to be fed to rabbits as they will husk the oats themselves, but winnowing must be carried out for other stock. To winnow, take the oats outside on a windy day and shake them from a bucket onto a tarp while standing on a stepladder. The chaff will blow away and the heavier oats will fall onto the tarp. Be sure it is not too windy or you will be sowing your domestic oats wastefully. *Yield:* 30–50 bushels per acre.

Soybeans

Soybeans are very high in protein but are somewhat less palatable than other stock feeds. They are grown very much like garden beans but require a longer growing season. Plant 6 inches apart in rows that are 3 feet apart. Allow the bean to mature fully so it is hard. Soybean can be fed whole if the animal will eat them, or ground as a component of a feed mix.

Sunflower Seeds

Sunflower seeds are a high-protein seed that are quite easy to grow and make excellent food for chickens. Sow the seeds where you want these tall plants to grow, allowing one foot between each plant. Let the flowers mature and "go to seed." When the seeds are dry, remove them from the flower head by rubbing it face down on ¹/₂-inch hardware cloth. The seeds can be fed whole as chickens shell them themselves.

Wheat

Get a variety of wheat earmarked for livestock feed purposes. Wheat can be sown by hand, as depicted in the famous painting by Millet, *The Sower*, or by use of a seed broadcaster, which can be purchased cheaply and will spread seed evenly as you turn the crank and walk down the plot.

Harvest is similar to oats, and threshing and winnowing are also necessary.

Storage

Grains can be stored in barrels or other covered containers free from both rodents and moisture. Make sure the seeds are dry enough (10–15¾ moisture content) before storing, or they may get moldy.

FORAGE CROPS

Pasture

Pasture, properly managed and of adequate size, can provide the most time- and labor-efficient way of feeding livestock. There are few regions where pasture alone can provide year-round forage for livestock.

Even in the heart of Dixie it is difficult to get more than 10 months of grazing out of well-managed pasture. In the northern reaches of the United States, 6 months is a reasonable length of the pasture season. In the heat of summer, most pastures are of minimal value unless the acreage available is large and the number of animals grazing per acre is low. Midsummer droughts and extreme heat are contributing factors to the poor growth of pasture grasses.

Earlier in the book it was mentioned that most grasses and legumes have preferred growing seasons. Those plant characteristics must be considered when developing a grazing plan so as not to overgraze and seriously damage the plants making up the pasture.

Pasture should be managed like any other crop. Periodic analysis of the soil is essential if one is going to return the soil those nutrients that have been taken out by the plants that have grown since the last fertilization. Your local agricultural extension agent can give you information regarding a soil test and where to send it (usually your state university).

Do not attempt to fool Mother Nature by planting crops that are not adapted to your region. Your results will probably be poor at best. Stick to varieties that have proved to be successful in your area, and leave the experimenting to those better able to afford it.

The secret, if there is one, to producing high-quality hay is to cut it early and insist on 100 percent cooperation from Mother Nature. For years I have looked at my mowing machine as a rain-maker, a hangover from some earlier tribal ritual. It can be the driest spring ever but once the mower is started and a piece of potentially beautiful hay has been mowed, the monsoon season will follow immediately.

Harvesting

Grass, hay, and legumes should be harvested as early as possible in order to ensure the highest level of nutrients, especially protein. Suggesting this practice and achieving it are two distinctly different things.

We are great late June, early July haymakers, well after the peak of nutrition has passed but when the weather is more suitable for haymaking. Commercial operations try to harvest their first crop of hay as haylage or grass silage where they are less dependent on the vagaries of the weather. In New England, the ideal time is late May or early June when the feed value of the crop is at its peak.

Most part-time farmers have time limitations that prevent them from synchronizing their available time, often weekends, and the desired amount of sunshine. The two never seem to mesh nicely, and to my knowledge the formula for doing so has yet to be devised.

As nostalgic as it might be to revert to a slower, more leisurely pace of doing this, the realities of our lives necessitate getting the haying done in as rapid and efficient manner as possible.

Buying Machinery

In most rural and semirural communities a variety of used machinery can sometimes be purchased for a reasonable cost. The purchase of a small, older model tractor, simple in design, can start you on your way to doing some of the necessary work yourself. When investing in used machinery, consider your own inclinations carefully. Are you happiest when trying to encourage an elderly piece of farm equipment to do its thing one more time, or does such activity work you into a stage of smouldering rage? Careful now, beating on that old baler with a ball-peen hammer probably isn't going to correct its malfunctioning parts.

So, if satisfaction derived out of keeping old stuff humming isn't your cup of tea, it's better to hire an enterprising young man with a custom service to do the work for you. This way you must pay, but you are going to pay one way or another. Take your choice.

Grazing

Don't overgraze your pastures. If you have animals that will graze very close, such as sheep and goats, have other pastures so that you can rotate them and let the others regrow. Don't put your stock to pasture too early in the spring, or it will never get a good start. Check with a local feed store, nursery, or your extension agent to determine what pasture crops are most suited to your locale and particular livestock needs. Good pastures (and hays) consist of a legume/grass mixture. The legumes supply protein while the grasses are high in energy. Also, the grasses will hold the soil better and prevent erosion and fill in well between the less dense legumes.

Kale and collards make good feed. These plants can be fed to poultry, sheep, rabbits, goats, pigs—most any livestock. They are very high in vitamins and minerals and can even be fed to young animals as they do not cause bloat the way many other succulent green forages may.

Fig. 8.1: Most Abundant Forage
Grasses and Legumes by Region

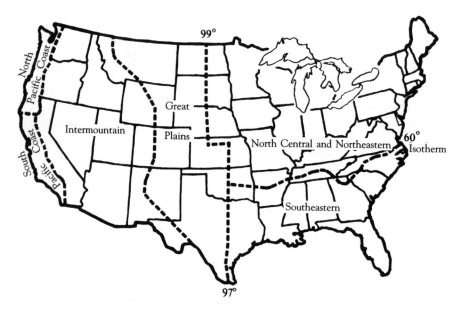

NORTH CENTRAL AND NORTHEAST

Grasses	Legumes
Kentucky bluegrass	white clover
timothy	Korean lespedeza
redtop	sweetclovers
orchard	alfalfa
Canada Bluegrass	common lespedeza
tall oatgrass	red clover
meadow fescue	alsike clover
smooth brome	hop clover
sudan	black medic
ryegrass	crimson clover
bents	birdsfoot trefoil
beardgrasses and	
bluestems	

SOUTHERN

Grasses	Legumes
Bermuda	common lespedeza
carpet	hop clover
dallis	white clover
vasey	Persian clover
redtop	black medic
Bahia	spotted burclover
fescue	
Rhodes	

GREAT PLAINS

Grasses	Legumes
grama	sweetclover
wheatgrass	alfalfa
buffalo	
brome	
galleta, tobosa, curly	
mesquite	
bluegrass, bluestem	
sudan	
bluegrass	
timothy	
redtop	
fescue	
Rhodes	

INTERMOUNTAIN

Grasses	Legumes
wheatgrass	alfalfa
grama	white clover
brome	sweetclover
galleta, tobosa, curly	alsike clover
mesquite	red clover
	black medic

SOUTH PACIFIC COAST

Grasses	Legumes
fescue	California burclover
brome	alfalfa
wild oats	white clover
Bermuda	black medic
sudan	strawberry clover
bluegrass	
timothy	
redtop	
Bermuda	
beardgrass or	
bluestem	

NORTH PACIFIC COAST

Grasses	Legumes
ryegrass	red clover
Kentucky bluegrass	white clover
bent	hop clover
red canary	alsike clover
orchard	
meadow fescue	
redtop	
timothy	
tall oatgrass	
meadow foxtail	

Source: USDA *Yearbook of Agriculture*, 1939.

Hay

Hay is your pasture cut, dried to about a 15 percent moisture content, and stored, either baled or loose, for winter feeding (or whenever pasture is scarce or not available). A legume/grass hay mixture is best, for the reason outlined above. But this is not to say that you must plow up your fields and reseed them. Recent studies have shown that a well-fertilized field of grass, properly cut and cured, will have nearly the same nutritional value as legume/grass hay. This again may make it cheaper in the long run, because legumes generally have to be reseeded more often than grasses.

Second-cutting hays (because they are more tender and have less hard stems) make the best feed and less is wasted. Most of the nutrients are concentrated in the leafy portions of the plants so hay that is cut when the plants are most leafy and are from well-fertilized fields offer the most nutrition. Table 8.1 illustrates the importance of an early cut and proper curing. It can be used as a guide both when making your own hay or when choosing hay that is offered for sale.

Table 8.1: How To Determine Hay Quality—Cornell Study

(1) *Stage of growth when cut* (the more mature, the more loss of nutritive value)

	Digestible Protein	Energy as T.D.N.
a. vegetative stage	18.7%	70%
b. bud stage	14.5%	63%
c. bloom stage	10.2%	56%
d. mature stage	6.4%	49%

(2) *Leafiness*—important (more proteins and vitamins in leaves)

(3) *Green color*—very important
  a. need 35%–60% of original color
  b. lose color—lose 90% vitamin A
  c. one or two rains on dry hay causes loss of 40%–60% feed value

(4) Compare alfalfa, a legume hay, and timothy, a grass hay, as to feeding value. Notice that the later grasses are cut, the lower the protein content.

	Digestible Protein	Energy as T.D.N.
a. early bloom timothy	4.2%	51.6%
b. full bloom timothy	3.2%	48.0%
c. late bloom timothy	2.4%	47.5%
a. leafy alfalfa	12.1%	51.1%
b. good alfalfa	10.3%	51.1%
3. stemmy alfalfa	8.2%	47.5%

Source: *Dairy Goats, Breeding/Feeding/Management.* New England Dairy Goat Industry Leaflet 439, Amherst: University of Massachusetts.

Silage

Silage is such crops as grasses, legumes, corn (plants as well as ears) and the like, chopped, stored, and fermented in the absence of air. After the silage is packed into the silo or other airtight container, the oxygen is used up in a short time and anaerobic bacteria produce lactic and other acids. As the acids increase, the bacteria die off; and, as long as no more air is permitted to enter, the silage will keep almost indefinitely.

If you choose to incorporate it into your feeding program, silage often can be purchased reasonably from neighboring farmers who may have a surplus for sale. On a very small scale, the aggravation of putting up silage is hardly worth it. If you want to give it a try, it can be done.

While the large silos attached to barns are most common for making silage, these are much too large for the small-scale operation we are talking about. Since the most important consideration in making silage is the absence of air, any reasonably airtight container can be used. Fifty-gallon drums are a good size and the most available.

Since silage will begin to spoil as you use it, the use of a number of drums will enable you to use one container at a time and not risk spoiling all your crop. Of course, any other airtight container, smaller or larger, can be used to suit your needs. If you really get into making silage and find it affords you good feed savings, you might want to construct a larger, more permanent silo. A large circular silo (square containers are less often used because of the difficulty of eliminating all trapped air from right-angled corners) made out of corrugated steel roofing fitted with a tight top and set on a concrete slab can be made easily and cheaply. If you place an airtight door at the bottom, you can remove the silage with a shovel without worry of spoilage.

Whether you use drums or other silos, the procedure for making silage is the same. The moisture content of the material to be ensiled is very critical. It should be between 55 and 65 percent moisture (as a guide, freshly cut hay is about 75 percent moisture); too moist and it will spoil easily; too dry and it will be unpalatable to your livestock. In order to pack and ferment well, the material should be chopped. Farmers use a silage chopper, but you will have to do this by hand or use a shredder or composter if you have one. It is best to have it cut into ¼ -inch lengths or be finely chopped. Pack it tightly into your silo and fit it with a tight cover. Every day for a week, stomp on it and press down to exclude any trapped air. Do this until the mixture is settled. Be sure the top is on tight and you should have a minimum of spoilage. Some people add molasses as a preservative but this is not essential. Your silage should be ready in about three weeks.

You will always have a little spoiling where air has leaked in after you begin removing the silage. Be sure to remove any moldy or spoiled silage before feeding it, and check your batch each time you remove some to be sure there isn't any undue spoilage.

Seeds and Fertilizing

Buy your seeds locally; these will be most adapted to your particular growing region. Check with local feed stores and, more important, your local extension agent, to determine what crops and varieties are best suited for your region.

Have your soil tested before sowing your crops and periodically every few years after to determine what is needed. Most state universities test soil samples for a nominal fee. They will recommend the correct amounts of chemical or organic fertilizers, depending on which you state as your preference.

Appendixes

Appendix A. Table of Helpful Facts

Animal	Female	Male	Young Animal	Castrated Male	Normal Temp. °F.
Chicken	Hen	Rooster	Chick[1]	Capon[2] Stag[3]	107.5°
Duck	Duck	Drake	Duckling	n.a.	(young: 102–106°)
Goose	Goose	Gander	Gosling	n.a.	
Turkey	Hen	Tom	Poult	n.a.	
Sheep	Ewe	Ram/ Buck	Lamb	Wether	100.9– 103.8
Goat	Doe	Buck	Kid	Wether	101–102°
Pig	Gilt[6] Sow[7]	Boar	Shoat/Piglet	Barrow[2] Stag[3]	101.6– 103.6°
Rabbit	Doe	Buck	Kindle	n.a.	102.5°
Calf	Heifer	Bull Calf	Calf	Steer	98–102.5°

Animal	Age of Puberty	Incidence of Heat	Duration of Heat	Gestation	Avg. Productive Life*
Chicken	4–6 mos.	n.a.	n.a.	21 days	2+ yrs.
Duck	5–7 mos.	n.a.	n.a.	28 days[4]	3+ yrs.
Goose	5–7 mos.	n.a.	n.a.	29–31 days[5]	3+ yrs.
Turkey	n.a.	n.a.	n.a.	28 days	3+ yrs.
Sheep	9 mos.[†]	13–19 days (avg. 16½)	3–72 hrs.	144–155 days (avg. 147)	10–12 yrs.
Goat	7 mos.	18–24 days (avg. 21)	1–2 days	145–155 days (avg. 150)	10–12 yrs.
Pig	3–5 mos.	16–24 days (avg. 21)	1–3 days	114 days	8–9 yrs.
Rabbit	6–7 mos.[8]	See text	See text	29–35 days (avg. 31)	3–4 yrs.
Calf	n.a.	n.a.	n.a.	n.a.	n.a.

*This is an *average* period of *good* production. This is not to say a ewe will not give birth to a live lamb after the age of 12, or a goat will definitely not milk after the age of 12; rather, these are guidelines to help you determine how many productive years an animal will give you on the average. In the case of poultry, while they will lay after the age of 2 or 3 years, their egg production will have decreased so markedly as to make them uneconomical, even for the backyard farmer.

[†]See Text

[1]Female: pullet; Male: cockerel
[2]Castrated before sexual maturity
[3]Castrated after sexual maturity
[4]Muscovy 35 days
[5]Canadian and Egyptian 35 days
[6]Before having a litter
[7]After having a litter
[8]Medium-weight rabbit breeds

Appendix B. Disease Tables

It is not the purpose of this summary to list every disease of each of the species described in this book, but rather to give the reader a brief overview of the more common diseases that affect the species discussed.

POULTRY

Infectious Disease	Poultry affected	Cause	Symptoms	Treatment	Prevention
Brooder pneumonia	Young chicken and turkeys especially	Fungus: *Aspergillus fumigatus*	Rapid breathing, labored, depression	None	Strict sanitation in brooder equipment
Blackhead or histomoniasis	Turkeys especially	Protozoa *Histomas melegridis*	Listless, ruffled feathers, dark blue head	Furazolidone, carbasone, dimethridazol, plus others	Strict sanitation; rotate ground; keep other birds away
Botulism Lembunich	Ducks	Bacteria: *Clostridium botulinum* (toxin)	Sudden death	None	Sanitation
Coccidiosis	All species	Coccidia, one or more variety for each species of bird	Bloody dropping in chickens	Amprolium, sulfonamides; get the current recommendations	Sanitation
Duck plague	Chicks, geese	herpes virus	Sudden death		Keep wild fowl away from flock and vaccination

POULTRY

Infectious Disease	Poultry affected	Cause	Symptoms	Treatment	Prevention
Fowl cholera	All	Bacteria *Pasteurella multocida*	Rapid death	Sulfa-quinoxaline	Vaccination
Fowl pox	Chickens, turkeys	Virus	Poxlike sores on unfeathered area of body	None	Vaccination in areas where disease is a problem
Viral hepatitis of ducks DVH	Ducks under 7 weeks	Pecorna virus	Sudden death	None	Rat and wild duck control; strict isolation of the young ducks; vaccination of adult breeder stock from older birds
Hemorrhagic enteritis of turkeys	Young turkeys	Group II adenovirus	Sudden onset, bloody diarrhea	None	Vaccination (in drinking water)
Infectious bronchitis	Chickens	Corona virus	Coughing, tearing	None	Good nursing, vaccination
Infectious coryza	Chickens	Bacterium *Hemophilus gallinarum*	Facial swelling may close eyes, discharge	Erythromycin	Buy replacement from clean flock; a bacterin is available
Infectious laryngotracheitis	Chickens	Virus	Gasping, coughing with hock extended	None	Vaccination
Influenza	All	Virus	Signs are those of a cold	None	Reportable disease

POULTRY

Infectious Disease	Poultry affected	Cause	Symptoms	Treatment	Prevention
Infectious serositis, "new duck disease"	Primarily ducks	Bacteria: Pasteurella antipestifer	Eye and nose discharge	Sulfaquin-oxaline, penicillin, steratomycin	Bacterin available
Newcastle disease	All	Virus	Coughing, sneezing, with nervous system involved in severe cases	None	Vaccination
Pullorum disease	Chickens, turkeys	Bacteria Salmonella pullorum	Symptoms in young ducks: listless, feces vent, do not eat	None	Control by testing of breeding flocks; irradiation program; vaccination
Marek's disease	Chickens	Virus. One of the first cancers proved to be caused by a virus	Paralysis, sometimes only depression	None	Vaccination

RABBITS

Disease	Cause	Symptoms	Treatment	Prevention
Viral Diseases				
Infectious myxomatosis	Virus poxvirus limited to West Coast	Milky eye discharge; ear edema; nasal discharge	None	None
Bacterial Diseases				
Pasteurellosis "Snuffles"	Pasteurella multocida	Thin pussy nose and eye discharge	None	Recovered animals may become carriers

RABBITS

Disease	Cause	Symptoms	Treatment	Prevention
Abscesses	Staphylococci, streptococci	Abscess on any part of body from fighting	None	
Mastitis (Blue breasts)	Staphylococci, streptococci	Hot, swollen mammary glands	Penicillin	Good sanitation
Treponematosis, vent disease	Spirochette: *Treponema cuniculi*	Scabs and/or loss of fur in genital region	Penicillin entire herd	Do not use affected animals for breeders
Hutch burn, urine burn	Wide range of possible organisms	Similar parts affected as vent disease	Antibiotic ointment; apply to affected parts	Keep hutches clean and dry
Enterotoxemia	*Clostridium spiroforme*	Sudden death	Onset usually too rapid to treat	Avoid diets too low in fiber
Mucoid enteropathy	Unknown	Constipation	None	Oral fluids to overcome dehydration
Tyzzus disease	*Bacillus piliformis*	Severe diarrhea	None	None
Parasitic Diseases				
Coccidiosis: hepatic and intestinal	Coccidia	Young rabbits may be off feed, dull	Sulfa-quinoxoline in feed or water	Good sanitation; avoid fecal contamination
Ear mites	Ear mite	Scratching at ear	Clean ear with peroxide then use ear mite medicine	Good sanitary practices
Miscellaneous Conditions				
Wet dewlap (moist dermatitis)	Various	Inflammation of the dewlap	Clip affected area—use antiseptic powder	Use automatic dew drop valves; elevate drinking dishes

RABBITS

Disease	Cause	Symptoms	Treatment	Prevention
Hair chewing	Perhaps low-fiber diet	Chewing of fur	Change to higher fiber diet	High fiber diet, e.g., good-quality hay
Hairballs	Swallowing fur	Indigestion, gagging	None	High fiber diet—mineral oil, etc. of no value
Heat exhaustion	Overexposure to bright sunlight	Heavy panting, elevated body temperature, and extreme lassitude.	Immerse severely affected rabbits in cold water	Construct hutches to allow for good ventilation. Provide cold water and sprinkle hutches in hot weather.
Sore hocks (ulcerative pododermatitis)	Direct contact with wire mesh, especially in heavy breeds where there is an accumulation of urine-soaked feces.	Raw, weeping sores on hind legs	Difficult to treat. Affected animals should be culled.	Keep hutches clean
Malocclusion	An inherited characteristic	Failure of teeth to grind against each other, especially the incisors or front teeth.	Affected teeth should be clipped off.	Affected animals should not be used as breeders.

SHEEP

Metabolic Diseases
Those caused by feed intake or stresses affecting all the body systems due to situations such as lambing.

Disease	Cause	Symptoms	Treatment	Prevention
Grass staggers, grass tetany, hypomagnesemic tetany	Low intake of magnesium; suddenly placing animals on lush pasture in the spring	Erratic behavior after being put on lush pasture	None	Magnesium supplements to fertilize pastures; feed dry hay before turning out initially
Parturient paresis	Unknown	Sudden onset 6–10 weeks prior to lambing especially in heavy ewes: muscle tremors; stilted gait; down; death	Calcium solutions I.V. or under skin	Good dietary management
Photosensitization	Light colored skin that has a hyperactive reaction to sunlight	Reddening and fluid (edema) in the skin.	Graze at night; steroids	Affected animals should not be used as breeders

Some breeds of sheep (Southdowns and Corriedales) have a defect existing in liver metabolism that prevents substances that predispose the animal to photosensitization from being eliminated in the usual way. Sheep of these breeds showing this trait should not be used as breeders.

Disease	Cause	Symptoms	Treatment	Prevention
White muscle disease	Selenium deficiency	Stiff gait, arched back, unable to arise	Selenium injections	Selenium supplement in feed

SHEEP

Disease	Cause	Symptoms	Treatment	Prevention
Pregnancy toxemia, ovine ketosis	Any factor that disrupts feed intake in fat; ewes carrying twins late in pregnancy	Nervous signs, inability to get on their feet; coma	Propylene glycol, steroids induce abortions, glucose	Induce exercise; feed away from shelter; avoid stressful situations where possible
Infectious Bacterial Diseases				
Tetanus	Bacteria: *Clostridium tetani*	Often follows surgical procedures in early life; "sawhorse" appearance	None	Administer tetanus antitoxin; later give toxoid.
Overeating disease	*Clostridium* types C, D	Sudden death, especially in weaned lambs on full feed	No effective treatment	Bacterin to those on full feed
Viral Diseases				
Bluetongue	Bluetongue virus spread by biting flies	Difficult breathing; reddening of muzzle, lips, ears; ulcer, erosions of mouth	None	Vaccine in affected areas only
OPP (ovine progressive pneumonia)	Retrovirus	Very slowly progressing disease in mature animals; slow, wasting disease	None	Test and slaughter program; isolate lambs at birth and feed milk from known negative ewes
Sore mouth, contagious ecthyma	Poxvirus	Lesions on lips, sometimes feet.	None	Recovered sheep highly resistant; vaccine available

SHEEP

Disease	Cause	Symptoms	Treatment	Prevention
Scrapie	Prion (structure smaller than a virus)	Loss of wool, intense itching, altered gait; loss of condition in sheep usually over 2 years	None	Test and slaughter
"Black disease" (infectious necrotic hepatitis)	Clostridium novyi	Sudden death	Toxin produced by organism together with liver flukes which have already damaged the liver	Toxoid (vaccine) in fluke region
Mastitis	Often Streptococcus or Staphylococcus species	Hot, swollen udder; abnormal secretion	None	Antibiotics; hot packs; frequent stripping
Strawberry foot rot	Fungus (actinomycete) Dermatophilus congalensis	Sores in lower leg	Penicillin	Self-limiting
Scours	Escherichia coli	Profuse diarrhea in newborn	Fluids to combat dehydration; antibiotics, antitoxemia drugs	Dip navels in iodine; strict sanitation in lambing pens
Joint ill	Erysipelothis/rhusiopathiae	Swollen joints and sometimes navel	Penicillin	Dip navels, strict cleanliness when castrating or docking

Parasitic Diseases

There are a wide range of parasites that affect sheep. Some invade various parts of the digestive tract while lungworms settle in the lungs. (These are largely covered in the text.)

GOATS

Goats are subject to much the same range of diseases as are sheep. The management of the various conditions follows the lines followed in sheep.

Disease	Cause	Symptoms	Treatment	Prevention
CAE (caprine arthritis and encephalitis)	Lentivirus	Encephalitis in young goats, arthritis in adults	None	Butazolidin in arthritic form; remove kids from doe immediately after birth and feed pasteurized colostrium.
Abscesses	Conpubacterium pseudotuberculosi	Abscess formation around head and neck.	None	
Urinary calculi (esp. in pet wethers)	Metabolic imbalance	Urinate with difficulty or not at all	Surgery to create an artificial opening	Feed diet with calcium phosphores ratio of 2:1; add ammonium chloride to diets; keep magnesium level low
Mastitis	Several different bacteria	Abnormal milk ranging from water to bloody	Variety of antibiotic preparations infused into udder; may need to run antibiotic sensitivity test to determine the best choice	Strict sanitation during the milking process; clean bedding and general environment

PIGS

Disease	Cause	Symptoms	Treatment	Prevention
Metabolic and nutritional iron deficiency	Milk naturally deficient in iron	Symptoms of anemia	None	Administer iron to newborn by injection; keep sod in pen
Hypoglycemia	Any factor that limits piglets' milk intake during first week of life especially	Low body temperature; listless; unresponsive	Give glucose intraperitonally	Be sure sow has enough milk; that all piglets are getting their share; keep warm and dry
MMA (lactation failure in sows)	Not understood at this time, a complex interaction of many factors	Piglets show signs of starvation; piglets depressed; increased temperatures; sows listless; refuse to let piglets nurse	Corticosteroids and broad spectrum antibiotics	Good management practices; place piglets on foster sow if available

Bacteria, Spirochetes, and Viruses

Disease	Cause	Symptoms	Treatment	Prevention
Bacterial brucellosis (a disease transmitted to man especially when handling infected carcasses at slaughter)	Brucella suis	Abortion: temporary or permanent sterility: lameness	None	Test and slaughter
Atrophic rhinitis	A complex disease with many contributing factors	Twisted snout may be end result	Several drugs effective including trimethoprim, tylocin, and tetracycline	Keeping a closed herd; good sanitation

PIGS

Disease	Cause	Symptoms	Treatment	Prevention
Enteric colibacillosis	Escherichia coli	Profuse watery diarrhea in young pigs	Restore fluid level; antibiotics as determined by sensitivity, previous experience	Avoid chilling and dampness; vaccinations of sows to increase level of necessary antibodies
Enteritis	Clostridium perfringens Type C	Bloody diarrhea in piglets, 1–5 days	None	Vaccination of pregnant sows is of some value to increase value of colostrum in affording protection
Edema	Escherichia coli	Disease of young pigs 5–14 days after weaning	Rapid death; wobbly; paddling of legs; head twist	Gradual change of feed from creep feed to weaning ration
Erysipelas	Erysipelothrix rhusiopathiae	Several different forms of this disease: (1) sudden death; high temperature, walking stiffly or remain lying down; (2) may show skin discolorations as "diamond skin disease"; (3) arthritis; (4) heart valve involvement	Penicillin	Vaccination
Leptospirosis (may act as reservoir to cause infection in man)	Leptospira pomona	Abortions; being off feed; listlessness	None	Vaccinations

PIGS

Disease	Cause	Symptoms	Treatment	Prevention
Mycoplasma pneumonia	*Mycoplasma hyopneumoniae*	Coughing, mild pneumonia; greatest problems: poor feed utilization and growth rate	Tylocin, Tetracycline	Establishment of disease-free herds
Pleuropneumonia	*Haemophilus pleuropneumoniae*	May be very severe with sudden deaths especially in young pigs; extremely difficult breathing; high fatality rate	Several antibiotics have been suggested including Tylocin and Trimethoprim	All-in, all-out management (all pigs in a group introduced simultaneously into area where they will be raised and later removed simultaneously).
Salmonellosis	*Salmonella cholerasuis*	Nursing pigs have diarrhea but die from septicemia; older pigs have bloody diarrhea	Some antibiotics may be of value but no drastic changes may be seen	Intensive clean-up after outbreaks
Swine dysentary	*Treponema hyodysenteriae* (a spirochete)	A mucous, bloody diarrhea the most common symptom	Do antibiotic sensitivity test and determine best antibiotic, which might include bacitracin, lincomycin, and others	Good sanitation treatment of carrier pigs
Tuberculosis (a disease transmissable from animal to man)	*Mycobacterium* species. Pig is susceptible to 3 types: cattle, bird, and man	Lymph gland involvement: a wasting disease	None	Good management (e.g. cleanliness, good ventilation)

PIGS

Viruses

Disease	Cause	Symptoms	Treatment	Prevention
Hog cholera (no longer present in U.S.A.)	DNA virus	Lethargy; off feed: high temperature	Hyperimmune serum	Vaccinations
Pseudorabies, "mad itch"	Herpes virus. The pig acts as a reservoir for the virus, which is highly fatal in other species such as cattle and sheep	In baby pigs, may see fever, trembling, convulsions; symptoms become progressively less severe as pig grows older	None	Bring in replacements from disease-free herds; segregation; isolation
Swine influenza, hog flu	Type A influenza virus	Rapidly spread throughout entire herd; high temperature; off feed; coughing; prostration	None	Good management with a stress-free environment
TGE (transmissible gastroenteritis)	Coronavirus	Vomiting followed by profuse, watery diarrhea; highly fatal in very young pigs, much less so in older pigs	None	Vaccinations

Parasites

Those worms affecting the stomach and intestinal tract of pigs can be controlled by the same types of medications used in other species. There are some worms peculiar to swine that should be mentioned.

Disease	Cause	Symptoms	Treatment	Prevention
Kidney worm infection	Stephanurus dentatus	Pigs tend to be "poor doers"; economic loss from contamination of affected organs and tissues	Ivermectin	"Gilts only" breeding program

PIGS

Disease	Cause	Symptoms	Treatment	Prevention
Trichinosis	*Trichinella spiralis*	Primarily a public health problem where infected, insufficiently cooked meat is eaten	None	Garbage containing pork should be cooked at 212°F for 30 minutes before being fed to swine

CALVES

Disease	Cause	Symptoms	Treatment	Prevention
Colibacillosis	*Escherichia coli*	Diarrhea; dehydration; weakness; death	Electrolytes, fluids, antibiotics	Buy calves from reputable sources; avoid stress; keep isolated; draft free; well-ventilated housing (calf hutch excellent)
Pneumonia	Various bacteria and viruses	Cough; difficulty breathing; off feed; fever	Fluids, antibiotics	

Appendix C.
A Homemade Incubator

While you can buy an incubator for hatching eggs, a homemade one is so simple to make that it's silly not to go that route. The only thing you have to purchase is a thermostat and possibly three lamp sockets and a few feet of wire.

CONSTRUCTION

Build a wooden box, either of plywood or scrap lumber, of the dimensions indicated. The depth of the box is twelve inches. Drill a ½-inch vent hole at a height of 1 inch above the egg drawer in each of the four sides (including the door)

and two holes in the top. The egg drawer should be built as indicated and slid into the incubator between the supports on each side so that it can be pulled in or out to check the eggs and turn them.

The window in front will enable those interested to watch the hatching process without constantly opening the door and lowering the inside temperature. The square piece of glass can be taped on over the hole or glued on, using a good epoxy. Hinge the door and use a hook and eye or other fastener to ensure that the door closes securely. The three light sockets should be mounted about 7 inches from the floor of the incubator and wired into the thermostat. Although I have not tried it with an incubator of this design, you might try wiring the center socket so that the light is constantly on. Then you will be able to watch the hatching without the thermostat cutting the lights off at the most exciting times.

The one bulb left on all the time should not keep the incubator too hot. If it does, replace the 25-watt bulb with a 15-watt bulb, and the incubator should operate within the desired temperature range.

The thermostat should be installed 2–3 inches from the level of the eggs in the egg drawer. It will then most accurately reflect the temperature of the eggs. A good unit for use in a homemade incubator can be purchased through a farm supply catalog or local hardware or feed store. The vent holes drilled into the sides and top of the incubator will ensure proper ventilation.

OPERATION

Turn your incubator on the day before you are ready to begin hatching the eggs so that it will heat up and set the thermostat at the proper temperature. Many thermostats don't have any calibrations, and you'll have to find the correct

Table C.1: Average Hatching Period for Common Fowl

Chickens	21 days
Geese	
Canada and Egyptian	35 days
All other breeds	29–31 days
Ducks	
Muscovy	35 days
All other breeds	28 days
Turkeys	28 days

temperature set by trial and error. You'll want to set the temperature within 99°–103° F. The temperature should remain between these extremes. Warm your eggs up to room temperature before placing them in the incubator so the temperature will not fall sharply when they are put it. (See Table C.1 for average times for hatching poultry eggs.)

The pan of water will help to maintain the proper humidity in the incubator. Keep it full at all times and do not permit the water to get stagnant. Finally, turn the eggs 4 times a day, at 6-hour intervals if possible. Make a *small* mark on one side of each egg to avoid confusion when turning.

Appendix D.
Tanning Skins

It is an unpardonable waste to throw the skins of your animals into the trash. If that is not enough, perhaps you will be convinced by the fact that you can make lovely and useful hats, gloves, and blankets from your tanned hides and even sell and barter them for additional income or goods. Sheepskins and rabbit skins first come to mind, but don't neglect calf skins or even your goat hides. The industrious person will find uses for all these skins.

Rabbit skins are perhaps the easiest to tan because of their size and because there is relatively little fat. Sheepskins are by far the most difficult (but you get the best finished product) because of their size and because of the large amount of fat that must be removed from the skin. Many taxidermists won't even consider tanning a sheepskin because of all the bother it can be. But that's because they scrape the fat off by hand, and that is a monumental job, believe me, because I've spent a day or two at it. It's a long and hard job and does not give very satisfactory results. The following methods will enable you to tan any type of skin (including sheepskins) with a minimum of time and effort.

PREPARATION OF SKINS

Use skins from freshly butchered animals. Those that have been dried or salted can be used as long as they are soaked well and are pliable before tanning. Wash skins in lukewarm water with a little detergent to remove loose dirt and bloodstains. Rinse well in cold water and wash as often as is necessary to remove all blood.

FLESHING

This why tanning has never caught on—fleshing *can* be terrible. With calf skins, rabbit skins, and other animal skins that don't have too much fat, you can flesh by hand in a few minutes and go right on to the next step, tanning. With sheep and goat skins, fleshing, or the removal of meat and fat from the skin, is a substantial job without the use of the pickle bath mentioned below. The following fleshing/

tanning bath is supplied by Jerome Belanger in his book *The Homesteader's Guide to Raising Small Livestock:*

Fleshing/Tanning Bath

8 fluid oz. battery acid or
2 fluid oz. sulfuric acid
2 lbs. (3 cups) pickling salt
2 gallons water

Mix enough of the solution to cover the skin and place it and the skin in a large stoneware crock or plastic container (don't put any acid in a metal container or it will corrode it). Weight the skin down so it won't rise out of the solution. A small hide may be done in less than a week, while a large sheepskin may take up to 2 weeks. You will learn how long with experience, but don't worry about damaging the pelts by keeping them in too long. They can stay in almost indefinitely as long as they are stirred in the solution periodically.

If the container is small and the skins are tightly packed, take them out every day and rearrange them so all sides get exposed to the tanning solution. In a larger container, or with smaller skins, take them out every 3 days, stir the mixture, and reimmerse them. The temperature of the solution is quite critical—keep it as close to 70° F. as possible. I tanned some skins one fall in our upstairs. The temperature fell into the fifties up there for a week, and the skins spoiled. Apparently it was too cool for the tanning to take place, but unfortunately not cool enough to retard the rotting process. The odor is minimal with this process, so you can keep it in your living area if the temperature elsewhere is too cool.

When finished, the hide should be taken from the solution and rinsed in cold water to remove any remnants of the acid. If all goes well, the flesh and fat will peel off the hide in large sheets. If you have ever fleshed a hide by hand, you will be amazed at how much easier this process is. What will take hours upon hours by hand will take perhaps 10 or 15 minutes after being in the acid bath. If the fleshing is still hard to complete, place the hide back in the solution for a few days until the fat peels off easily.

Once the fleshing is complete, rinse, and put back into the solution for another week. After this, wash in warm water and rinse well. I have got good results using this method alone, but if your results are not satisfactory (in regard to the quality of the leather and hide after tanning), you can go on to the tanning procedures, listed next. Again, I have found these steps unnecessary so if you try it and agree, skip the next section and go right on to finishing.

TANNING

After fleshing, either by hand or with the above process, you can tan using one of the following methods:

Alum Mixture

1 lb. alum
2 lbs. (3 cups) salt
Mix with enough water to cover skins

Dissolve the alum first in hot water, then add salt and enough cold water to achieve the correct solution. Keep the skin in this solution for 2 days and turn 3 times a day. (Alum, an astringent, can be purchased in most pharmacies).

Tannol Mixture

1 lb. Tannal
2½ lbs. (3¾ cups) salt
10 gallons water

First use an acid bath as explained above. Then keep hide in Tannol solution for 5 days and stir 3 times a day. (Tannol is an aluminum sulfate compound which can be purchased from taxidermy supply companies.)

FINISHING

If you want to remove the hair from your hide, you can pull it out after it's been either in the acid bath or in one of the tanning solutions (the hide must still be wet for the hair to come out easily).

After removing the hide from whichever solutions you use, rinse the skin in cold water to remove any solution that may be left. Then let the skin drip dry out of sunlight. Before the skin is completely dry and is still pliable, the leather side may be oiled. Then you must work the leather to make the hard, brown skin soft and white. Constant stretching, pulling, and rubbing over a smooth board will do the trick in time. Be sure to do this work before the skin dries out. Another easy way to finish the hide is first to stretch it and nail it to a frame (any holes or cuts in the hide should be sewn up before stretching to prevent further ripping of the hide), and then sand it with an electric sander or by hand. After the leather is softened, comb burrs and other chaff out of the fleece with a metal comb, trim the skin to even it up (I usually cut the legs off a hide), and you're all set.

Appendix E. Manure Table

Animal	Weight[1]	Manure[2] per Animal per Day[3] (lbs.)	per Animal per Year (lbs.)[3]	% Urine	% Feces	% Nitrogen	% Phosphorus	% Potassium
Chicken	5	.26	95	n.a.	n.a.	1.4	0.5	0.6
Sheep and Goat	100	4.0	1,460	50	50	1.1	0.2	0.8
Pig	500	32.5	11,860	45	55	0.7	0.2	0.5
Rabbit	4	.09	34	50	50	2.4	1.4	0.6
Calf	100	6.0	2,190	50	50	0.6	0.2	0.4

[1]This is an average figure. For chickens, sheep and goats, and pigs this is an average for an adult. For the rest of the animals it is a point ¾ between birth weight and slaughtering weight.

[2]Raw manure. That is, urine and feces with no bedding.

[3]Based on average weight from Column 1.

Note: There can be up to a 20 percent variation in the amount of manure depending on feed, environment, and health factors. Urine, pound for pound, is of greater fertilizing value than feces (except in pigs), so good absorbent bedding is most important. Manure that is left outside will have less value than that which is under some cover.

Remember also that much of the manure will be voided on pasture. To determine how much you can save, figure the number of days the animal is confined and multiply times the entry under pounds of manure per day. Similarly, for those animals that don't live a full year, figure the number of days alive and multiply times pounds per day.

Appendix F.
Catalog of Resources

BOOKS AND OTHER PUBLICATIONS

Note: A number of out-of-print books and publications have been included in this bibliography, as they remain useful references for readers looking for additional information on particular topics. Some may be found in your local library. The vocational agriculture department of your local high school or technical school might also have some of these publications. For readers located near a state university with a college of agriculture, the college library would be a good source of in- and out-of-print publications.

Most publications of state extension services and the U.S. Department of Agriculture (USDA) are available free, or at minimal cost, from your local office of the Cooperative Extension Service. Consult your telephone directory for the office nearest you. For further information on this type of publication see "Other Sources of Information" following this section.

General

Atherton, H.V., and W.A. Dodge. *Guide to Milk Sanitation Terms*. Burlington: University of Vermont Extension Service Brieflet 1076.

Battaglia, R.A., and V.B. Mayrose. *Handbook of Livestock Management Technics*. Minneapolis, MN: Burgess, CEPCO Division, 1981.

Belanger, Jerome. *The Homesteader's Handbook to Raising Small Livestock*. Emmaus, PA: Rodale Press, 1974.

Bundy, Clarence R., et al. *Livestock and Poultry Production*. Englewood Cliffs, NJ: Prentice-Hall, 1975.

Ensminger, M.E., and R.O Parker. *Sheep and Goat Science*. Danville, IL: Interstate Printers and Publishers, 5th ed., 1986.

Haynes, N. Bruce. *Keeping Livestock Healthy: A Veterinary Guide*. Charlotte, VT: Garden Way, 1978.

Iowa State University. *Livestock Waste Facilities Handbook*. Ames: Iowa State University, Midwest Planning Service, 2nd ed., 1985.

Merck and Co., Inc. *The Merck Veterinary Manual.* Rahway, NJ: Merck and Co., 1986.

U.S. Department of Agriculture. *Raising Livestock on Small Farms.* Washington, DC: USDA Farmer's Bulletin 2224.

Wilson, J.V. *Weaning Is for Anyone.* New York: Van Nostrand Reinhold, 1975.

Ziegler, P.T. *The Meat We Eat.* Danville, IL: Interstate Printers and Publishers, 12th ed., 1985.

Poultry

Banks, Stuart. *The Complete Handbook of Poultry Keeping.* New York: Van Nostrand Reinhold, 1979.

Graves, Will. *Raising Poultry Successfully.* Charlotte, VT: Williamson, 1985.

Mercia, Leonard. *The Family Laying Flock.* Burlington: University of Vermont Brieflet 1218.

_____. *Raising Poultry the Modern Way.* Charlotte, VT: Garden Way, 1975.

Robinson, Ed. *Producing Eggs and Chickens with the Minimum of Purchased Feed.* Charlotte, VT: Garden Way, 1972.

University of Vermont. *Culling for Higher Egg Production.* Burlington: University of Vermont Agricultural Extension Service Circular 115V.

U.S. Department of Agriculture. *Raising Ducks.* Washington, DC: USDA Farmer's Bulletin 2215.

_____. *Raising Geese.* Washington, DC: USDA Farmers Bulletin 2251.

_____. *Small Poultry Flocks.* Washington, DC: USDA Farmers Bulletin 2262.

Rabbits

Bennett, Bob. *Raising Rabbits the Modern Way.* Charlotte, VT: Garden Way, 1975.

_____. *Raising Rabbits Successfully.* Charlotte, VT: Williamson, 1984.

Cheeke, Peter R. *Rabbit Feeding and Production.* Orlando, FL: Academic Press, 1987.

Cheeke, P.R., N.M. Patton, and G.S. Templeton. *Rabbit Production.* Danville, IL: Interstate Printers and Publishers, 1982.

Choate, Peter. *Rabbit Production.* Charlotte, VT: Williamson, 1987.

Kanable, Ann. *Raising Rabbits.* Emmaus, PA: Rodale Press, 1977, out of print.

Pennsylvania State University. *Producing Rabbits Without Harmful Residues.* State Park: The Pennsylvania State University Cooperative Extension Service.

Sanford, J.C. *The Domestic Rabbit.* Somerset, NJ: Wiley, 1957, out of print.

Templeton, George. *Domestic Rabbit Production.* Danville, IL: Interstate Printers and Publishers, 1968, out of print.

Thear, Katie. *Practical Rabbit-Keeping.* Woodstock, NY: Beekman, 1981.

U.S. Department of Agriculture. *Selecting and Raising Rabbits.* Washington, DC: USDA Agricultural Information Bulletin 358.

_____. *Commercial Rabbit Raising.* Washington, DC: USDA Agricultural Handbook 309.

Sheep

Colby, Byron E. *Hints on Feeding and Managing Ewes and Lambs.* Amherst: University of Massachusetts Extension Service.

_____. *Suggestions on the Economical Feeding of Sheep.* Amherst: University of Massachusetts Extension Service.

_____. *Well Fed Sheep Are Profitable Sheep.* Amherst: University of Massachusetts Extension Service.

Juergenson, E.M. *Approved Practices in Sheep Production.* Danville, IL: Interstate Printers and Publishers, 1981.

Kruesi, William K. *The Sheep Raiser's Manual.* Charlotte, VT: Williamson, 1985.

Midwest Plan Service. *Sheep Housing and Equipment Handbook.* Ames, IA: Midwest Plan Service, 3rd ed., 1982.

Nielson, G.R. *Control of Lice and Ticks on Sheep.* Burlington: University of Vermont Extension Service.

Nova Scotia Department of Agriculture. *Home Tanning of Sheepskins.* Halifax, NS, Canada.

Sheep Industry Development Program, Inc. *The Sheep Production Handbook.* Englewood, CO: American Sheep Industry Association, rev. ed., 1990.

Simmons, Paula. *Raising Sheep the Modern Way.* Pownal, VT: Storey Communications, rev. ed., 1989.

Ricketts, G.E., R.D. Scoggins, and D.L. Thomas. *Recommendations for a Sheep Management Program.* Urbana-Champagne, IL: University of Illinois, College of Agriculture, Cooperative Extension Service Circular 1221, 1983.

University of Connecticut. *Keeping the Newborn Lamb Alive.* Storrs: University of Connecticut Department of Animal Husbandry.

_____. *Lambing Troubles.* Storrs: University of Connecticut Department of Animal Husbandry.

Goats

American Dairy Goat Association. *Dairy Goat Breeding—Management.* Spindale, NC: American Dairy Goat Association. (Available from: Agricultural Publications, University of Connecticut, U–36, 1376 Storrs Road, Storrs, CT 06269–4035.) Cooperative Extension Services of the Northeastern States.

Belanger, Jerry. *Raising Milk Goats the Modern Way.* Charlotte, VT: Garden Way, 1975.

Haenblain, G.F.W, and D.L. Ace. *Extension Goat Handbook.* Washington, DC: USDA.

Halliday, Jill and John. *Practical Goatkeeping.* Woodstock, NY: Beekman, 1990.

Luttman, Gail. *Raising Milk Goats Successfully.* Charlotte, VT: Williamson, 1986.

University of Massachusetts. *Dairy Goats: Breeding/Feeding/Management.* Amherst: University of Massachusetts Leaflet 439.

U.S. Department of Agriculture. *A Dairy Goat for Home Milk Production.* Washington, DC: USDA Leaflet 538.

Pigs

Albers Milling Co. *Swine.* A Carnation-Albers publication.

Belanger, Jerome. *Raising the Homestead Pig.* Emmaus, PA: Rodale Press, 1977, out of print.

Kellogg, Kathy and Bob. *Raising Pigs Successfully.* Charlotte, VT: Williamson, 1985.

Krider, J.L., J.H. Conrad, and W.E. Carroll. *Swine Production.* Heightstown, NJ: McGraw-Hill, 1982, out of print.

Pennsylvania State University. *The Pennsylvania State Swine Management Handbook.* University Park: Pennsylvania State University College of Agriculture.

Pond, W.G., and J.H. Maner. *Swine Production and Nutrition.* New York: AVI, 1984.

U.S. Department of Agriculture. *Breeds of Swine.* Washington, DC: USDA Farmers Bulletin 1263.

van Loon, Dirk. *Small-Scale Pig Raising.* Charlotte, VT: Garden Way, 1978.

Veal

Albers Milling Co. *Calf Plan.* A Carnation-Albers publication.

Beef

Grohman, Joann S. *Keeping a Family Cow.* New York: Ballantine, 1984.

Hobson, Phyllis. *Raising a Calf for Beef.* Charlotte, VT: Garden Way, 1976.

van Loon, Dirk. *The Family Cow.* Charlotte, VT: Garden Way, 1975.

Feeds and Feed Crops

Cullison, A.E., and R.S. Lawrey. *Feeds and Feeding.* Reston, VA: Reston Publishing, 3rd ed., 1981.

Langer, Richard. *Grow it!* New York: Avon, 1973, out of print.

Miller, D.A. *Forage Crops.* Heightstown, NJ: McGraw-Hill, 1984.

Morrison, Frank B. *Feeds and Feeding.* Ithaca, NY: Morrison Publishing Co., 23rd ed., 1967. (Although long out of print, this text has long been considered the bible of books on feeds and feeding.)

Rodale, J.I., et al. *The Encyclopedia of Organic Gardening.* Emmaus, PA: Rodale Press, 1975.

OTHER SOURCES OF INFORMATION

The federal government's thousands of publications offer a source of material often left untapped because of the difficulty of finding what you want. Try presenting your need to your Extension Service office or the state or district offices of your congressman or senator.

Your local county extension agent will offer free advice on raising livestock and growing crops. Many excellent publications, some of them free, are available through the Extension Service. Write to:

Extension Service Director,
c/o College of Agriculture,
Your State University

On a national level, you can address your needs to:

Superintendent of Documents
Government Printing Office
Washington, DC 20402
(202-783-3238)

The Government Printing Office has 24 bookstores scattered around the country.

Consumer Information Center
Box 100
Pueblo, CO 81002

Government and Publications Office
USDA
Washington, DC 20584

The U.S. Department of Agriculture offers many publications, some of them free, on raising livestock. Each year the USDA publishes a yearbook on a selected topic relating to agriculture. These yearbooks are available, free of charge, by writing to your senator or congressman.

Pennsylvania State University Publications has many publications that will be helpful to the beginning livestock grower. Write to:

Publications Distribution Center,
112 Agricultural Administration Building,
University Park, PA 16802

Free publications include:

5165	Feeding the Pig
5252	Preventing Swine Disease Problems
5303	Calf and Heifer Housing
S313	Management of Gastrointestinal Parasites in Sheep
EC354	Guide to Swine Nutrition
EC358	Farming Management
VM188	Home Processing of Poultry
VM191	Suggested Feeding Programs for Small Poultry Flocks

PERIODICALS

American Agriculturalist Inc.
710 West Clinton St.
P.O. Box 516
Ithaca, NY 14851
Sub: $8/year in United States (monthly).

American Farmland
American Farmland Trust
1920 N St. NW, Suite 400
Washington, DC 20036
(Non-profit organization devoted to preservation of American Farmland) $15 annual dues.

American Journal of Alternative Agriculture
Institute for Alternative Agriculture
9200 Edmonston Road
Suite 117
Greenbelt, MD 20770
Sub: for individuals $20/year
Journal devoted to promoting organic farming (quarterly).

Farmers Digest
Box 624
Brookfield, WI 53008-0624
Sub: $12/year; 10 issues/year

Farm Journal
Farm Journal Inc.
230 West Washington Square
Philadelphia, PA 19105
Sub: $12; 14 issues/year
For families who own/operate farms and ranches.

Countryside and Small Stock Journal
Countryside Publications Ltd.
W8333 Doepke Rd.
Waterlook, WI 53594
Sub: $18/year
A magazine devoted to country living.

Keystone Dairy Goat News
For information write:
Verna Guzel
R.D. #1 Box 39A
Canonsburg, PA 15317

Progressive Farmer
Box 2581
Birmingham, AL 35202
Sub: $12/year; monthly.

Successful Farming
Meredith Corp.
1716 Locust Street
Des Moines, IA 50336
Sub: $12/year; 12 issues/year. All subscription orders must show farm or ranch
connection as owner or operator or in an occupation directly related to
agriculture.

Wallaces Farmer
Farm Progress Companies, Inc.
1501 42nd Street
Suite 501
West Des Moines, IA 50265
Sub: $20/year outside Iowa; $12/year in Iowa.
Subscriber must show a direct tie to agriculture.

Rural Heritage
P.O. Box 516
Albin, IA 52531
Sub: $12/year; 4 issues a year. A magazine devoted to showing how the rural past
is alive today.

Sheep Magazine
Box 329, 120 South Highland
Jefferson, WI 53549
Sub: $15/year; 8 issues a year

Sheep Breeder and Sheepman Magazine
Mead Livestock Services
P.O. Box 796
Columbia, MO 65205
Sub: $18/year

Small Farmer's Journal
Small Farmer's Journal, Inc.
P.O. Box 2805
Eugene, OR 97402
Sub: $17/year; 4 issues/year

Periodicals are published by the various colleges of agriculture at state universities. Check with your state university to see what is published there.

Some excellent farm magazines are circulated on a regional basis. A check with your county extension office should give you information regarding such publications in your area.

PAPERS AND NEWSPAPERS

Farmshine Newspaper
P.O. Box 372
Lititz, PA 17543
Serving PA and neighboring states
Sub: $8/year; weekly except Christmas week

Farm and Home News
Lee Publications
Grand Street
P.O. Box 121
Palatine Bridge, NY 13428
Weekly; write for subscription rates

Lancaster Farming
P.O. Box 609
1 East Main Street
Ephrata, PA 17522
Sub: $12.50/year; weekly.

The Shepherd
Sheep and Farm Life, Inc.
5696 Johnston Road
New Washington, OH 44854
Published monthly. $14

SOURCES OF SUPPLIES

Here is a list of sources of supplies, from farm machinery to day-old chicks. All provide catalogs; some companies charge for them.

Forester Supplies
205 W. Ranlain St.
PO Box 8397
Jackson, MS 39204
Forestry, farming equipment
Free catalog

Morden Farm and Ranch
1825 Big Horn Road
Cody, WY 82414
Farm and ranch supplies
Free catalog

Midstates Wool Grower Coop.
8270 Nieman Road
Shawnee Mission, KS 66214
Livestock supplies
Free catalog

A.M. Leonard
6665 Spiker Road
Piqua, OH 45356
Gardening and farming equipment
Free catalog

NASCO
901 Jamesville Ave.
Ft. Atkinson, WI 53538
Farm and ranch supplies
Free catalog

Stockman's National Supply
250 S. Santa Fe Ave.
Pueblo, CO 81003
Farm and ranch equipment
Free catalog

Surplus Center
1015 West O St.
Lincoln, NE 68501
Farm machines, engines
Free catalog

Murray McMurray Hatchery
Webster City, IA 50595
Chickens
Free catalog

Pilgrim Goose Hatchery
Williamsfield, OH 44093
Ducks and chickens
50 cents for catalog

Strombugs
Pine River, MM 56474
Chickens, poultry equipment
$1.00 for catalog

Country Hatchery
PO Box 747
Wewoka, OK 74884
Live poultry
Free catalog

Grain Belt Hatchery
PO Box 125
Windsor, MO 65360
Exotic, popular chickens
50 cents for catalog

Allen Hatchery
Box 46–9
Windsor, MO 65366
Chickens
50 cents for catalog

Country General Store
103 N. Monroe
Waterloo, IA 53594
"Back to nature" equipment
Free catalog

Reich Poultry Farms
Box 14K
Marietta, PA 17547
Chickens
Free catalog

Glossary

Balling gun. An instrument that is slipped partly down the throat of an animal and used to administer large pills.

Barrow. A male pig castrated before sexual maturity.

Boar. A male pig.

Bolus gun. See Balling gun.

Broiler. A chicken of either sex that weighs 2½ pounds and is less than 8 months old (cf. Fryer and Roaster).

Broken mouth. A condition occurring in sheep and goats, usually at about the age of 5–6 years, whereby some of the permanent teeth are missing from the mouth (cf. Full mouth, Gummer).

Brooder. The enclosure in which young animals are kept, or are allowed free access to, which furnishes heat until they adapt to outside temperatures.

Broodiness. A natural condition in poultry in which a female goes out of production and sets on and attempts to hatch eggs.

Broody coop. An enclosure used to confine a broody hen and hasten her return to normal production.

Buck. A male goat. A male rabbit. Less commonly: a male sheep.

Capon. A male chicken castrated surgically before sexual maturity (cf. Stag and Caponette). In butchering, a castrated male chicken that weighs 6–8 pounds.

Caponette. A male chicken that is neutered before sexual maturity by the implantation of female hormones (cf. Capon).

Carbonaceous hay. See Grass hay.

Castrate. To remove the testicles of a male animal.

Chaff. Unwanted parts of grain separated during winnowing.

Chevon. Goat meat.

Chick. A young chicken.

Closed-faced. A sheep that has considerable wool covering about the face and eyes. This often leads to a condition known as wool blindness (cf. Open-faced; also Wool blindness).

Cock. Any male chicken butchered after 8 months of age. A stag.

Cockerel. A male chicken less than 1 year old.

Colostrum. The milk produced by an animal immediately and for the first few days after the birth of its young. Rich in antibodies that protect the animal against many diseases. A young animal that does not receive this milk may die or will be very difficult to raise.

Creep feeding. The feeding of a young animal by means of an enclosure accessible to it, but not to its mother.

Crop. A digestive organ in poultry in which food is prepared for digestion.

Crossbred. An offspring that results from the breeding of 2 purebred parents of different breeds. A hybrid.

Crossbreeding. The mating of purebred parents from different breeds (cf. Hybrid vigor).

Cull. (v.) To remove an inferior animal from a flock. (n.) Any animal that is culled.

Dicalcium phosphate. A mineral mix rich in calcium and phosphorus.

Disbud. To remove the horns of an animal.

Dock. (v.) To cut short the tail of an animal (most commonly a lamb); usually for sanitary reasons and to facilitate breeding in females. (n.) The area around the tail of sheep or other animals.

Doe. A female goat. A female rabbit.

Drake. A male duck.

Dress out. To remove the feathers or skin and to cut up and trim the carcass of an animal after slaughter.

Drylot. An area of confinement containing little or no natural feed into which all or most of an animal's feed must be brought (cf. Free-ranging).

Duck. A female duck.

Duckling. A young duck of either sex.

Elastrator. A tool used in castrating and docking. A tight rubber band is applied to the tail or the scrotum; the circulation is thereby cut off and the tail or scrotum gradually dries up and falls off.

Emasculatome. A tool used for docking and castrating that both cuts and crushes surrounding tissue to prevent excessive bleeding.

Ewe. A female sheep.

Eyeing. Clipping the wool from around the face of closed-faced sheep to prevent wool blindness.

Farrow. To give birth to a litter of piglets.

Feed conversion ratio. The rate at which an animal converts feed to meat. If an animal requires 4 pounds of feed to gain 1 pound it is said to have a four to one (4:1) feed conversion ratio.

Field crops. Feed plants grown primarily for their seeds. For example, corn, wheat, oats, soybeans (cf. Forage crops).

Finishing. The increased feeding of an animal just prior to butchering, which results in rapid gains and increased carcass quality.

Flail. (n.) a tool used for threshing grain. (v.) To use a flail to thresh grain.

Flushing. The practice of increasing the feed intake of a female animal just prior to ovulation and breeding. This causes the animal to gain some weight and drop more eggs, often resulting in larger litters.

Forage crops. Those plants or parts of plants that are used for feed before maturing or developing seeds (cf. field crops). The most common forage crop is simple pasture.

Fowl. For butchering purposes, a female chicken that is more than 8 months old.

Free choice feeding. A type of feeding routine whereby feed, water, and salt are provided in unlimited quantities and an animal is left to regulate its intake (cf. Hand feeding).

Free-ranging. Allowing animals, especially poultry, to roam freely and eat as they wish without any sort of confinement (cf. Drylot).

Freshen. To come into milk, as when a dairy animal gives birth.

Fryer. A chicken of either sex that weighs $2\frac{1}{2}$–$3\frac{1}{2}$ pounds and is less than 8 months old (cf. broiler and roaster).

Full mouth. A state in sheep or goats when an animal has a full set of permanent teeth. This occurs at approximately the age of 4 years. The animal will continue to have what is known as a full mouth until it loses some teeth (cf. Broken mouth) or until it loses all of its teeth (cf. Gummer).

Gander. A male goose.

Garbage. Scraps. Leftover food. A very poor choice of words as it has bad connotations.

Gilt. A female pig that has not yet produced a litter (cf. Sow).

Girth. The measure of the distance around an animal at a point just behind the front legs.

Gizzard. The muscular enlargement of the alimentary canal of poultry that immediately follows the crop; has thick, muscular walls and a tough horny lining for the grinding of the food.

Goose. A female goose.

Gosling. A young goose of either sex.

Grade. An animal of no distinguishable breed or background (cf. Purebred).

Grading up. The practice of improving a flock whereby purebred sires are mated to grade animals and their offspring. In 3 generations the offspring will be ⅞ purebred and in some cases eligible for registration. Upgrading.

Grass. Any of the members of the family *Gramineae* (such as timothy, orchardgrass, Sudagrass, etc.). When used as pasture, hay or in silage, they provide more energy than legumes but are lower in protein and vitamins (cf. Legume).

Grass hay. Any hay totally or primarily from a grass crop (cf. Legume hay; Grass; Legume).

Grass lamb. A lamb that is dropped in the springtime and is raised on pasture in the summer months and butchered in the fall when pasture dies.

Grit. Any ingested rough, undigestible matter such as bits of glass or small stones that is used to grind food in the gizzard of a chicken or other poultry.

Gummer. A sheep or goat having no teeth. (cf. Full mouth, Broken mouth).

Hand feeding. A type of feeding routine whereby an animal is fed measured amounts of food, water, salt, etc. (usually just food) at fixed intervals (cf. Free choice feeding).

Hay. Any crop (most often grasses or legumes) that is cut, dried, and stored (either baled or loose) for later use.

Heifer. A cow that is under 3 years of age and has not yet produced a calf.

Hen. A female chicken. A female turkey.

Heterosis. See Hybrid vigor.

Hothouse lamb. A lamb that is dropped in the fall or early winter and is marketed at an age of 6–12 weeks. These are usually sold in periods before the Christmas or Easter holidays to take advantage of higher prices.

Hybrid. See crossbred.

Hybrid vigor. The increase of size, speed of growth, and vitality of a crossbreed over its parents. Heterosis.

Inbreeding. The mating of very closely related animals such as mother and son, father and daughter, brother and sister. In experienced hands it can be used to selectively maintain certain desirable traits; if used improperly, it can produce undesirable traits and downgrade stock (cf. Linebreeding).

Kid. (v.) To give birth to a young goat. (n.) A young goat.

Kindle. (v.) To give birth to a litter of rabbits. (n.) A young rabbit.

Lamb. (n.) A young sheep of any sex, less than 1 year old. (cf. Mutton, Yearling). (v.) To give birth to a lamb (or lambs).

Lambing Loop. A length of smooth or plastic-coated wire used as an aid in difficult lambing.

Legume. Any of the members of the family *Leguminosae* (such as clovers, alfalfa, trefoil, vetches). When used as pasture, hay, or in silage, they provide more protein and vitamins than a comparable grass crop (see Grass).

Legume hay. Any hay made totally or primarily from a legume crop (cf. carbonaceous hay; Grass hay; Legume).

Linebreeding. Very similar to inbreeding but the breeding is not so close, for example, the mating of cousins (cf. Inbreeding).

Milk replacer. A powder that when mixed with water is fed to young animals as the milk portion of their diet.

Molt. To shed feathers, fur, skin, or horns and replace them with new growth.

Mutton. In butchering, any sheep over 18 months of age (cf. Lamb, Yearling).

Needle teeth. The 8 sharp teeth present in newborn piglets. Especially in large litters, these teeth can cause injury to other piglets and the sow's udder and should be clipped (see text).

Open-faced. Sheep that naturally have little or no wool covering about the face and eyes. This is a desirable trait as it discourages the problem of wool blindness. (See Wool blindness; cf. Closed-faced.)

Piglet. A young pig of either sex.

Pinning. The sticking of a young lamb's tail to its anus. This will prevent normal bowel action and result in constipation and, if not loosened in time, death.

Polled. Without horns; either naturally or by artificial means.

Poult. A young turkey of either sex.

Pullet. A young hen less than 1 year old.

Purebred. An animal of a recognized breed whose lineage has been kept pure (i.e., not mixed with another breed) for many generations (cf. Grade).

Purebreeding. The mating of purebred parents from the same breed (cf. Crossbreeding).

Ram. An uncastrated male sheep or goat.

Replacement animal. A young animal that is being raised to take the place of an older animal that is being culled; e.g., since a person wants to keep a flock of 5 breeding ewes, he will have to buy two *replacement ewes* to take the place of the two older breeders that have died.

Ringing. Clipping a breeding ram around the neck, belly, and penis in order to facilitate mating.

Roaster. A chicken of either sex that weighs 3½–5 pounds and is less than 8 months old (cf. Broiler and Fryer).

Rooster. A male chicken.

Ruminant. Any one of a class of animals including sheep, goats, and cows that have multiple stomachs. They are most efficient feeders because bacterial action in one of the stomachs, the rumen, increases the food value of low-grade food.

Scours. Technically, a bacterial infection in calves and sheep that results in a whitish-yellow, foul-smelling diarrhea. Informally, any diarrhea.

Scraps. Edible refuse that is saved from table scraps or collected from restaurants, food stores, etc. and fed to livestock.

Settle. To become pregnant.

Sex-linked. Distinguished sexually at birth, as chicks, by differences in coloration.

Shoat. See Piglet.

Silage. A feed consisting of certain roughages and/or field crops finely chopped, tightly packed in an airtight container, and allowed to ferment in the absence of air.

Sow. A female pig that has produced a litter (cf. Gilt).

Stag. A male chicken castrated after sexual maturity (cf. Capon). Also, any male chicken butchered after 8 months of age. A cock. A male pig castrated after sexual maturity.

Strip. To remove milk from the teat of an animal by sliding the fingers from the base of the teat to the end.

Swill. Scraps.

Tagging. Clipping the wool from around the dock of a ewe so that it does not interfere during mating.

Teasing. Keeping a ram in sight of, but not in contact with, ewes just prior to breeding. This often stimulates ovulation in the ewes.

Thresh. To separate grains from the plant, as in removing oats from straw.

Tom. A male turkey.

Uterine capsule. A medication administered to a ewe to prevent infection after entering her with a hand or other instrument during lambing.

Wattles. Fleshy appendages hanging from the neck of a chicken, turkey, or goat.

Wean. To remove a young animal from its mother and accustom it to food other than its mother's milk.

Wether. A castrated male lamb; a castrated male goat.

Windrow. Hay or any other crop raked into a row to dry.

Winnow. To remove chaff from grain by using a current of air.

Wolf teeth. Needle teeth.

Wool blindness. A condition that develops most often in closed-faced sheep due to irritation of the eyes by wool and particles of chaff contained therein (see Closed-faced and Open-faced).

Yearling. Any animal aged one year to eighteen months. (In sheep, cf. Mutton and Lamb.)

Index

Books from The Countryman Press

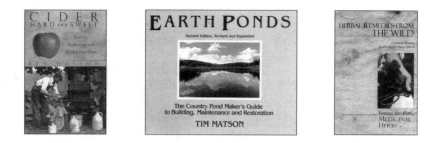

Country Living

Backyard Sugarin'

Cider, Hard & Sweet: History, Traditions, and Making Your Own

Earth Ponds: The Country Pond Maker's Guide

*The Earth Ponds Sourcebook: The Pond Owner's Manual and Resource
    Guide*

The Earth Shall Blossom: Shaker Herbs and Gardening

Herbal Remedies from the Wild: Finding and Using Medicinal Herb's

Living with Herbs: A Treasury of Useful Plants for the Home and Garden

The New Complete Guide to Beekeeping

*The New England Herb Gardener: Yankee Wisdom for North American Herb
    Growers and Users*

Perennials for the Backyard Gardener

*The Granite Kiss: Traditions and Techniques of Building New England
    Stone Walls*

Cookbooks

The King Arthur Flour 200th Anniversary Cookbook

Seasoned with Grace: My Generation of Shaker Cooking

Wild Game Cookery: Down-Home Recipes for Food from the Wild

We offer many more books on hiking, bicycling, paddling, travel, nature, and other subjects. Our books are available at bookstores and outdoor stores everywhere. For more information or a free catalog, please call 1-800-245-4151, or write to us at The Countryman Press, P.O. Box 748, Woodstock, Vermont 05091. You can find us on the Internet at www.countrymanpress.com